*FM 3-100.38
*MCRP 3-17.2B
*NTTP 3-02.4.1
*AFTTP(I) 3-2.12

FM 3-100.38 — U.S. Army Training and Doctrine Command, Fort Monroe, VA

MCRP 3-17.2B — Marine Corps Combat Development Command, Quantico, VA

NTTP 3-02.4.1 (REV A) — Navy Warfare Development Command, Newport, RI

AFTTP(I) 3-2.12 — Headquarters Air Force Doctrine Center, Maxwell Air Force Base, AL

23 August 2001

Procedures for Unexploded Ordnance Operations

TABLE OF CONTENTS

Page

*This publication supersedes FM 100-38, MCRP 4-5.1, NWP TP 3-02.4.1, ACCPAM 10-752, PACAFPAM 10-752, and USAFEPAM 10-752, 10 July 1996.

FIGURES

TABLES

PREFACE

1. Scope

This unclassified multiservice tactics, techniques, and procedures (MTTP) publication describes the unexploded explosive ordnance (UXO) threat and provides guidelines to minimize the impact of UXO hazards. It provides warfighting personnel at the operational and tactical levels with information to optimize UXO safety and increase efficiency, while reducing or eliminating losses of personnel and equipment to UXO hazards.

2. Purpose

This MTTP provides commanders and their units guidelines and strategies for operating with UXO hazards, while minimizing the impact of these hazards on friendly operations. This publication will facilitate coordination, integration, and force protection requirements regarding UXO during joint exercises, contingencies, or other operations. It also identifies functions and responsibilities for planning, reporting, tracking, and clearing UXO to accomplish the mission, while minimizing the impact of UXO on the operation.

3. Applicability

This publication applies to all commanders, leaders, staff, and warfighters participating in joint operations. This publication will assist the joint force commander (JFC), service component commanders, their staffs, and anyone responsible for force protection of personnel and resources.

4. Implementation Plan

Participating service command offices of primary responsibility will review this publication, validate the information, and reference and incorporate it in service and command manuals, regulations, and curricula as follows:

Army. The Army will incorporate the tactics, techniques and procedures (TTP) in this publication in United States (U.S.) Army training and doctrine publications as directed by the Commander, U.S. Army Training and Doctrine Command. Distribution is in accordance with Department of the Army (DA) Form 12-99-R.

Marine Corps. The Marine Corps will incorporate the TTP in this publication in U.S. Marine Corps (USMC) training and doctrine publications as directed by the Commanding General, USMC Combat Development Command. Distribution is in accordance with the Marine Corps Publication Distribution System.

Navy. The Navy will incorporate the TTP in U.S. Naval doctrine and training publications as directed by the Commander, U.S. Navy Warfare Development Command. Distribution is in accordance with the Navy military standard requisitioning and issue procedure (MILSTRIP) desk guide and Navy Supplemental Publication 409.

Air Force. U.S. Air Force (USAF) units will validate and incorporate appropriate procedures in accordance with applicable governing directives as validated by Headquarters (HQ), Air Force Civil Engineer (CE). Distribution is in accordance with Air Force Instruction 33-360 as directed by the Air Force Doctrine Center.

5. User Information

The Air Land Sea Application (ALSA) Center is the proponent for this publication with the joint participation of the approving service commands. ALSA will review and update this publication as required.

We encourage recommendations for changes to improve this publication. Key specific comments to the chapter, page, and paragraph. Also provide the rationale for each recommendation. Send comments through the appropriate service doctrine point of contact, for the ALSA Center.

EXECUTIVE SUMMARY
UXO
Multiservice Procedures for Unexploded Ordnance Operations

This manual—

- Describes the UXO threat to friendly forces and operations.
- Provides guidelines for planning, reporting, tracking, and marking UXO hazards and training recommendations for the joint force.
- Provides commanders several options for applying force protection measures against UXO hazards.
- Delineates the roles of explosive ordnance disposal (EOD) and engineer units with regard to UXO hazards.
- Describes service-specific UXO missions and capabilities.

Introduction

Every military operation can produce UXO hazards. UXO pose a threat to operational mobility, personnel, equipment and facilities. Understanding the impact of UXO hazards and the procedures used to minimize their effects can maximize the efficiency of resources available during a joint operation. Establishing procedures and training the force to react to these hazards enhances joint force capabilities. This publication contains TTP to mitigate UXO hazards while conducting joint operations.

Concept

This publication makes a critical distinction between land-based mines and other UXO hazards. Mines and minefields are, by definition, a category of UXO; distinction of the specific type of UXO is critical to determine which methods and forces to employ to mitigate the hazards of mines and other UXO. The breaching, reduction, or clearing of land-based mine hazards is primarily the responsibility of combat engineer units; the reduction or clearing of all other UXO hazards is primarily the responsibility of EOD units. (More details concerning this distinction are available in Chapter 2 of this publication.) Non-UXO trained military personnel should avoid, mark, and report UXO hazards using the methods prescribed by this publication. When possible, include UXO-trained personnel in any operation involving UXO. EOD technicians are the ordnance experts, however, EOD forces comprise less than one percent of the entire force. It is critical that EOD experts conduct detailed clearance of UXO hazards. This publication is not intended to train Soldiers, Marines, Sailors, or Airmen as UXO experts.

Organization

The chapters in this MTTP address UXO hazards, procedures for avoiding UXO hazards during joint operations, and unique service capabilities. The appendices of this MTTP are quick references defining specific skills needed when UXO hazards are encountered.

Chapters

Chapter I (UXO Hazards)—Defines UXO hazards and describes the degree of risk for different operational categories (such as maneuver, air assault, aviation, amphibious, and air base operations).

Chapter II (Joint Force UXO Operations)—Describes the responsibilities for planning and executing a joint operation with UXO hazards. This chapter introduces the UXO reporting, marking, and tracking requirements and defines considerations when planning and conducting operations with UXO hazards. It also defines options commanders may use when confronting UXO hazards.

Chapter III (Service Component Capabilities)—Outlines individual service missions, command and control (C2) structures, and specific capabilities of engineer and EOD forces of each service.

Appendices

Appendix A (Identifying UXO)—Assists non-UXO trained personnel in identifying the different types of UXO hazards. This appendix complements Appendix C by providing methods to describe and identify UXO hazards during reporting. Identification of the ordnance is a requirement for line 4 of the UXO Spot Report.

Appendix B (Marking UXO)—Describes the procedures for marking UXO hazards to warn personnel operating in the proximity of the hazard and assists clearance personnel in finding the hazard.

Appendix C (Reporting UXO)—Describes the routing and format of the nine-line UXO Spot Report.

Appendix D (Protective Measures)—Provides three methods to protect personnel and equipment when operating near UXO hazards: evacuate, isolate, or barricade. This appendix further describes the benefits of each method and provides examples to assist with implementation.

Appendix E (UXO Graphics)—Establishes graphic control measures to assist commanders, staff, and warfighters in tracking UXO hazards.

Appendix F (UXO and Obstacle Numbering)—Describes the numbering methodology for tracking and labeling UXO and obstacles on the engineer obstacle overlay.

Appendix G (UXO Training)—Provides leaders a guide for training military personnel on identifying, marking, and reporting UXO hazards. This appendix also describes the protective measures necessary to survive when operating in an environment with UXO hazards.

PROGRAM PARTICIPANTS

The following commands and agencies participated in the development of this publication:

Joint

Joint Staff, J-7, Joint Doctrine Education and Training Division, Washington, D.C.

Joint Staff, J-34, Combating Terrorism, Washington, D.C.

Department of Defense EOD Technology and Training Secretariat, Indian Head, MD

Joint Warfighting Center, Fort Monroe, VA

Commandant, Naval School EOD, Eglin AFB, FL

Army

HQDA, ATTN: DALO-AMA-EOD, Washington, D.C.

U.S. Army Training and Doctrine Command, Deputy Chief of Staff, Doctrine (ATTN: ATDO-A), Fort Monroe, VA

XVIII Airborne Corps, Assistant Corps Engineer, Fort Bragg, NC

HQ, 52nd Ordnance Group (EOD), Fort Gillem, GA

HQ, 79th Ordnance Battalion (EOD), Fort Sam Houston, TX

U.S. Army Technical Detachment, Navy EOD Technology Division, Indian Head, MD

Army EOD Training Representative, Fort Lee, VA

Marine Corps

Marine Corps Combat Development Command, Joint Doctrine Branch (C427) and Ground Branch (C422), Quantico, VA

Marine Corps Base, EOD, Quantico, VA

HQ, USMC Logistics, Planning and Operations, Washington, D.C.

Marine Corps Directorate, Navy EOD Technical Division, Indian Head, MD

Seventh Engineer Support Battalion, Camp Pendleton, CA

Eighth Engineer Support Battalion, Camp Lejeune, NC

Navy

Navy Warfare Development Command, ALSA Liaison Officer (LNO), Norfolk Naval Base, Norfolk, VA

Commander, EOD Group ONE, San Diego, CA

Commander, EOD Group TWO, Norfolk, VA

Air Force

HQ, USAF Civil Engineer, Washington, DC

Air Force Doctrine Center, Detachment 1, Langley AFB, VA

Air Combat Command, CE/EOD Division, Langley AFB, VA

Air Force Civil Engineer Support Agency, Tyndall AFB, FL

Air Force Special Operations Command, CE/EOD Division, Hurlburt Field, FL

Air Force Materiel Command, CE/EOD Division, Wright-Patterson AFB, OH

HQ, AF Space Command, CE/EOD Division, Peterson AFB, CO

HQ, Air Mobility Command, CE/EOD Division, Scott AFB, MO

Air Education Training Command, CE/EOD Division, Randolph AFB, TX

HQ, USAF Europe, Ramstein AFB, CE/EOD Division, Germany

HQ, Pacific Air Force, CE/EOD Division, Hickam AFB, HI

75th CE Group, Hill AFB, UT

56th CE Squadron, Luke AFB, AZ

Detachment 63, Aircraft and Armament Center, Indian Head, MD

Chapter I
UXO Hazards

1. Background

a. **Introduction.** Saturation with UXO has become a characteristic of the modern battlespace and will likely continue to threaten military forces and operations. U.S. personnel have been killed or injured by UXO in virtually every conflict or contingency in which the United States has participated. This can be attributed primarily to unfamiliarity with UXO countermeasures and avoidance procedures. Every individual participating in a joint operation should be able to recognize and react safely to UXO hazards. To this end, leaders should train their personnel to conduct operations in UXO contaminated environments. Commanders should further consider risks to personnel and operations from UXO and integrate the impact of the UXO hazard into mission planning. This MTTP provides the appropriate information for planning, implementing, and executing operations to minimize risks to forces and operations from UXO. Examples illustrated in this manual are U.S. ordnance; however, most foreign militaries possess explosive ordnance with similar capabilities and hazards.

b. **Joint Publication (JP) 1-02 Definitions.**

(1) UXO. Explosive ordnance that has been primed, fused, armed, or otherwise prepared for action, and fired, dropped, launched, projected or placed in such a manner as to constitute a hazard to operations, installations, personnel or materiel and remains unexploded either by malfunction or design or for any other cause.

(2) EOD. The detection, identification, on-site evaluation, rendering safe, recovery, and final disposal of UXO. It may also include explosive ordnance that has become hazardous by damage or deterioration.

2. Hazards

c. **Production of Hazards.** The United States prides itself on the low dud rates of its munitions; however, all explosive ordnance has the potential to become UXO. Therefore, commanders and staff should plan for duds and the impact on future maneuver or cleanup after hostilities are complete. The actual hazard area produced by UXO depends on the type and density. The following examples provide a method to predict the potential magnitude of a UXO hazard:

- A fire mission of 36 multiple launch rocket system (MLRS) rockets could produce 1,159 (36 X 644 X 5 percent dud rate) UXO hazards in the target area.
- A B-52 dropping a full load of 45 cluster bomb units (CBUs) (with each CBU containing up to 650 submunitions) may produce 1,462 (650 X 45 X 5 percent dud rate) UXO hazards.
- The family of scatterable mines (FASCAM) can produce a significant amount of UXO hazards. For example, the bomb live unit (BLU)-91 and 92/B Gator system can dispense hundreds of mines covering an average area of 200 by 650 meters in a matter of seconds. All scatterable mines have a self-destruct mechanism, ranging from four hours to 15 days; however,

if the self-destruct mechanism fails they could become UXO hazards. Other U.S. and foreign mines can be programmed to self-destruct up to 300 days from battlefield delivery.

a. **Impact on Operations.** UXO concerns all ground, air, and maritime forces because all forces operating in areas with UXO hazards are at risk. UXO affects operational and tactical planning and execution of operations. The presence of UXO in operational areas can add considerable time to any operation. Commanders and planners can minimize unnecessary delays by accurately planning for UXO hazards, while also having forces trained to respond correctly when these hazards are encountered. Without prior planning and coordination, maintaining the operational tempo is more difficult in a UXO environment.

b. **Degree of Risk.** Variables affecting the degree of risk include, but are not limited to—

(1) Types and density of explosive ordnance employed by enemy and friendly forces.

(3) Protection available to personnel, equipment and facilities (such as armored vehicles versus dismounted infantry, revetted aircraft over exposed aircraft, or fortified positions over exposed positions).

(4) Mission and degree of mobility required of the affected force.

(5) Terrain and climatic conditions.

c. **Impacts to Ground-Based Mobility.** UXO inhibits mobility by—

(2) Restricting the use of terrain, while reducing momentum (to include speed of maneuver and rates of march).

(6) Increasing reconnaissance requirements.

(7) Inhibiting night movement, while increasing risks.

(8) Increasing the risk to combat, combat support, and combat service support elements.

(9) Requiring the allocation of trained forces and other resources to clear and mark lanes.

(10) Reducing combat power due to potential losses of personnel and equipment.

d. **Hazards to Armored/Mechanized Forces.** Armored and mechanized vehicles offer varying degrees of protection from UXO. Mechanized crews can reduce their risk from crossing UXO hazard areas by remaining mounted; but UXO can disable the vehicles by breaking or throwing tracks. Armored and mechanized commanders must also consider the increased risk to non-armored vehicles and dismounted forces when operating in UXO environments. If crews dismount their vehicles, they increase their risk of injury or death from UXO hazards.

e. **Hazards to Dismounted Forces.** Dismounted forces face the greatest risk from UXO. To reduce this risk, dismounted forces need to be able to identify the type of UXO hazard correctly, understand procedures for avoiding and reporting these hazards, and be aware of UXO hazard areas previously reported.

f. **Hazards to Non-Armored Vehicles.** Personnel in non-armored vehicles face nearly the same risk from UXO as dismounted forces because the protection afforded by non-armored vehicles is negligible. The non-armored vehicles are also vulnerable to damage or destruction from UXO.

g. **Hazards to Air Assault and Aviation Forces.** Air assault and aviation forces are also at risk to UXO. Aircraft in defilade, flying nap of the earth, or in ground effect (below 45 feet) are vulnerable to UXO. U.S. and foreign munitions incorporate sensitive fusing that can react to aircraft in ground effect. Aviation units should be aware of hazard areas and conduct a thorough reconnaissance before occupying or conducting operations (to include occupying assembly areas and forward arming and refueling points).

h. **Hazards to Air Base Operations.** UXO contaminating aircraft launch, recovery, and support areas have the potential to affect expeditionary air forces. Mitigating these hazards before landing any forces and establishing operations requires adequate planning and allocation of necessary forces to accomplish the task.

i. **Hazards to Amphibious Landing Craft.** UXO has the potential to significantly damage certain types of landing craft and in some cases, result in casualties of embarked personnel. Sea and anti-landing mines can also deny access to selected landing sites and sea-lanes. Mitigating these hazards before landing forces requires adequate planning and allocation of necessary forces to accomplish the task.

j. **Hazards to Civilians and Civil Infrastructure.** UXO poses a significant hazard to civilians and the civilian infrastructure within the area of operations (AO). Because most civilians are not familiar with military ordnance, they are largely unaware of the extreme hazard that UXO presents. Due to this unfamiliarity, civilians, often-curious children, will pick up and handle UXO items with devastating effects. Civilians have also been killed or injured by buried UXO while performing routine tasks, such as walking or digging. In an effort to reclaim homes and reestablish lives during post-conflict operations, local civilians have collected and returned hazardous UXO items to U.S. forces. Unfortunately, due to the potentially large number of UXO and their long-term impact, these items will continue to be a deadly hazard to unwary indigenous personnel for decades after cessation of military operations. Additionally, the effects of UXO on the civil population and infrastructure can create significant, and often very undesirable, diplomatic, economic, and information impacts.

Chapter II
Joint Force UXO Operations

1. Purpose

a. **Introduction.** Joint operations have become the routine method of employing U.S. forces. This chapter provides standardized methods for joint force C2; EOD and engineer employment; reporting, marking, and tracking UXO; and protective measures when operating in a UXO environment. It also defines the command, staff, and warfighting responsibilities employed to minimize UXO risks to the joint force.

b. **Minefields and UXO Differentiation.** The joint term for UXO includes mines and minefields; however, there are basic differences in how a joint force manages mines and other UXO hazards. This MTTP will differentiate between mine hazards and all other UXO in specific circumstances. This distinction guides the commander's determination of which methods and forces to employ to mitigate the hazards based on the situation. The breaching, reduction, or clearing of land-based mine hazards is the responsibility primarily of Army and Marine Corps combat engineer units. The reduction or clearing of all other UXO hazards is the responsibility primarily of EOD units. In rapid operational tempo, high-intensity conflicts, UXO hazards may become mobility obstacles, especially in breaching operations and while establishing critical main supply routes. Combat engineers, or other non-EOD trained personnel, may then be called upon to reduce or clear non-mine UXO hazards, but only on a limited basis and under the direct technical guidance of EOD forces. Because of the greater risks involved when non-EOD trained soldiers are used to reduce UXO hazards, the commander must include—

- Thorough assessments of the risks and control measures available.
- Technical EOD personnel to supervise the execution.
- Detailed pre-execution training on common in-theater UXO hazards and safety procedures.

2. Operational Considerations

a. **General.** UXO hazards affect the morale of the fighting force, restrict mobility, increase logistical burdens, divert resources from the primary mission, and produce casualties. Integrated planning, reporting, and tracking of UXO throughout the theater are critical to successful joint and coalition operations. The air tasking order, rules of engagement, and coordinating instructions in the operations plan (OPLAN) or operations order (OPORD) are tools to inform the force and reduce the impact of UXO on joint forces.

b. **Considerations.** The following are operational and tactical considerations when operating in an environment with UXO:

(1) Availability, prioritization, and task organization of limited theater EOD and engineer forces.

(2) Impact on current and future military operations (including the terrain; effect on timing; positioning of forces; routes for maneuver and logistics units; potential nuclear, biological, and chemical [NBC] detection and decontamination support; and survivability requirements).

(3) Known UXO locations and densities (coalition and enemy).

(4) Increased logistical requirements (including barrier and demolition material).

(5) Method/process and communications infrastructures to report, track, and disseminate UXO hazards.

(6) Designating recovery, disposal, and safe holding areas.

(7) Security requirements.

(8) Impact on the civilian populace and infrastructure.

c. **Commander's Options.** There is no single device or technique that will eliminate every UXO encountered. Clearance operations are inherently hazardous and consume time. After assessing the UXO hazards germane to the mission, several options are available to the commander:

(1) Bypass the hazard.

(2) Employ tactical breaching procedures.

Note: Conventional minefield breaching methods (such as tank plow or hand-emplaced explosives) may be effective against limited types of UXO. Consult engineer or EOD personnel before proceeding, as the effectiveness is highly dependent on the unit's breaching capability and the specific type of UXO.

(3) Self-extract to alternate routes or positions.

(4) Restrict from use.

(5) Clearance by priority area.

(6) Complete clearance.

(7) Accept the risk of casualties and continue with the assigned mission.

3. Responsibilities

Responsibilities for planning, directing, managing, and executing the joint UXO mission extend from the combatant commander throughout the staff, to the subordinate JFC to the service and functional command forces. Appendices A through G provide guidelines to assist the entire force with UXO identification and marking, reporting, protecting, tracking, and training.

a. **JFC.** The JFC's and supporting combatant commander's guidance may address specific considerations for UXO hazards. The guidance may include priority of engineer and EOD support, management of joint fires, and C2 measures. The JFC's intent should provide safety guidance and establish anti-fratricide procedures within the joint operations area (JOA). The JFC should also include EOD and engineer special staff officers in all operational planning to take full advantage of their ordnance and mine clearing/breaching expertise.

b. **Intelligence Directorate of a Joint Force (J-2).** The J-2 is responsible for preparing and disseminating all known information on the enemy ordnance order of battle, weapon systems, activity that results in UXO hazard areas, minefields, and locations of ammunition supply points. This information must include known and suspected NBC capabilities and their potential employment.

c. **Operations Directorate of a Joint Staff (J-3)/Logistics Directorate of a Joint Staff (J-4).** The J-3/J-4 staff plans and coordinates the commander's guidance and establishes procedures to inform the joint force of all UXO hazard areas. The J-3, in coordination with the joint force engineer and EOD staffs, establishes joint force reporting and tracking procedures. There is often confusion whether the J-3 or J-4 should direct and manage EOD and engineer tasks. The solution is to ensure the OPLAN/OPORD clearly identifies J-3/J-4 responsibilities. A common method for the JFC's assignment of staff responsibility is to have the J-3 manage/direct combat engineer tasks, and the J-4 manage/direct EOD tasks. It is often more efficient for the JFC to direct the J-3 to manage immediate engineer and EOD tasks affecting land combat operations, and direct the J-4 to manage rear area engineer and EOD tasks.

d. **Plans Directorate of a Joint Staff (J-5).** The J-5 incorporates the commander's guidance regarding joint force UXO reporting and tracking procedures into OPLANs/OPORDs. The J-5 considers future terrain needs, while minimizing the impact of enemy and friendly-generated UXO. Additionally, the J-5 ensures plans and orders include adequate safety of personnel and anti-fratricide procedures.

e. **Joint Rear Area Coordinator (JRAC).** The JRAC plays a critical role in UXO reporting in the joint rear area (JRA) and must be part of the coordination and information network dealing with UXO. The JRAC is responsible for creating a secure environment in the JRA to facilitate sustainment, host nation support, infrastructure development, and joint force movements. The JRAC ensures that all personnel and units operating in the rear area follow JFC-established UXO hazard tracking, disseminating, and reporting procedures. The JRAC must also coordinate with the JFC staff, lead service (if designated), and/or joint task force (JTF) that has control of the EOD and engineer forces responsible for UXO operations in the JRA.

f. **Joint Force Engineer.** The joint force engineer is the staff element responsible for planning, reporting, and tracking minefields. The joint force engineer augments the J-2, J-3, J-4, and J-5 staff sections and advises the JFC on task organization requirements for minefield breaching and clearing operations. During operations, the joint force engineer receives, consolidates, and disseminates minefield information to the staff and affected units. The joint force engineer also incorporates minefield and UXO hazard area information onto the engineer obstacle overlay. The engineer obstacle overlay is the primary source depicting UXO hazard areas. The engineer staff maintains all reports and historical data on minefield and UXO hazard areas for use during future operations.

g. **Joint Force EOD Staff Officer.** The joint force EOD officer is the staff element responsible for coordinating the detection, identification, recovery, evaluation, safe rendering, and final disposition of explosive ordnance. He is typically assigned to the J-4 staff section, and augments the J-5 with planning issues relating to exploded ordnance. In addition to his coordinating staff responsibilities, he specifically—

(1) Establishes and operates an EOD-incident reporting system.

(2) Establishes, operates, and manages technical intelligence reporting procedures.

(3) Coordinates requirements for EOD support with service components, multinational partners, and civilian organizations, as appropriate.

h. **JFC Service Component Units.** Service component responsibilities include identification, marking, reporting, and tracking UXO hazards within their assigned AO. See Appendix A, *Identifying UXO*; Appendix B, *Marking UXO*; Appendix C, *Reporting UXO*; Appendix D, *Protective Measures*; Appendix E, *UXO Graphics*; Appendix F, *UXO and Obstacle Numbering*; and Appendix G, *UXO Training*.

i. **Joint EOD Task Force (JEODTF).** A JEODTF may be formed and jointly staffed as a subordinate JTF controlling (via operational control [OPCON] or tactical control [TACON] of attached units) two or more service component EOD organizations. The JEODTF commander is responsible for making recommendations to the senior JTF (or geographic combatant commander if a senior JTF has not been formed) on the proper employment of EOD and on how to accomplish assigned missions. See FM 4-30.16/MCRP 3-17.2C/NTTP 3-02.5/AFTTP(I) 3-2.32 for additional information on a JEODTF.

j. **Joint EOD Operations Center (JEODOC).** A JEODOC is a multifunctional operations center under the auspices of the JTF J-4. The primary purpose of the JEODOC is to manage theater-level UXO hazard-reduction operations and EOD planning, integrating, coordinating, and tasking function (through the direction and authority of the commander) when there is no subordinate JEODTF formed to accomplish this task. The JFC, with recommendations from the J-4, determines the need for the JEODOC whenever the EOD management requirements are beyond the capability of the J-4 or subordinate

EOD force headquarters. See FM 4-30.16/MCRP 3-17.2C/NTTP 3-02.5/AFTTP(I) 3-2.32 for additional information on a JEODOC.

4. Capabilities

a. **EOD.** Army, Marine Corps, Navy, and Air Force EOD units diagnose, render safe, and dispose of UXO that threaten personnel, facilities, and operations throughout the JOA. Joint regulations and Department of Defense (DOD) directives prescribe specific responsibilities for each service. (Chapter 3 of this publication provides specific roles and capabilities of each service.) Common EOD training, equipment, and technical manuals provide each service the capability to—

(1) Perform actions to locate and gain access to UXO.

(2) Conduct diagnostic actions to identify and evaluate UXO.

(3) Perform render safe procedures by applying special methods and tools.

(4) Direct actions to recover or perform final disposition of UXO.

(5) Perform technical intelligence collection and exploitation of first seen or previously unknown ordnance.

(6) Perform these procedures on U.S. and foreign (conventional or NBC) UXO and improvised explosive devices (IEDs).

b. **Army and Marine Corps Combat Engineers.** (See Chapter 3 of this publication for service-specific roles and capabilities of engineers.) Navy Seabees and Air Force civil engineers do not play a leading role in UXO clearance or breaching operations. Land combat operations will most likely involve countering enemy barriers, obstacles, and minefields. Army and Marine Corps combat engineers provide mobility, countermobility, and survivability in support of the OPLAN/OPORD. They conduct mine/countermine operations and obstacle breaching to support the commander's overall scheme of maneuver. Breaching operations make maneuver possible, regardless of enemy obstacle efforts. Since units can encounter obstacles when conducting any movement, joint forces integrate breaching operations into all movement plans. When possible, units bypass enemy minefields to maintain momentum and conserve critical countermobility assets. However, the joint force must conduct breaching operations when channelized into an enemy kill zone.

c. **Non-UXO Trained Personnel.** As previously discussed (this chapter, paragraph 1.b.), commanders use non-EOD personnel to reduce mobility obstacle-related UXO hazards in exceptional circumstances. When non-EOD personnel perform limited UXO hazard removal and destruction, these non-EOD personnel require special training and supervision by qualified EOD personnel. Normally, the recommended procedure for non-UXO trained personnel locating a UXO hazard is to avoid, mark, and report the hazard using the nine-line UXO Spot Report (see Appendix C). Appendices A through G provide guidance and assistance regarding

UXO identification, marking, reporting, tracking, protective measures required, and training.

5. Identifying

Recognizing a UXO hazard is the initial and most important step in reacting to a UXO hazard. Proper identification when reporting UXO assists explosive ordnance reconnaissance teams in understanding the type and characteristics of the hazard and the resources needed to clear or render the hazard safe (see Appendix A).

6. Marking

When mines and UXO are encountered, they should be clearly marked using the two different standard markers. The national stock number (NSN) 6230-00-926-4336 is the standard marker for mines, and NSN 7690-01-463-3422 is the standard marker for UXO. These standard markers warn other friendly personnel and non-combatants of the potential hazard (see Appendix B for proper marking procedures).

7. Reporting

a. **Introduction.** Timely and accurate UXO reporting and intelligence information gathering procedures are critical to the safe conduct of operations within the JOA. During joint operations, utilization of the standard UXO Spot Report will prevent redundancy and ensure accurate incident reporting, tasking, and tracking. An example of the standard UXO Spot Report is at Appendix C. Line 4 of the UXO Spot Report requires the identification of the ordnance. Appendix A provides useful guidelines and information to describe the type of hazard correctly.

b. **Dissemination.** Timely dissemination of the UXO Spot Report allows joint forces to counter the hazards associated with UXO effectively. The J-3 receives the UXO Spot Report through the chain of command and forwards the report to the joint force engineer. The joint force engineer converts the UXO Spot Report onto the engineer obstacle overlay for dissemination to subordinate units. The joint force engineer then processes, plots, and disseminates the information to higher, lower, and adjacent units.

8. Tracking

a. **Tracking Methods.** Appendices C, E, and F provide useful information to assist commanders and staff with the UXO tracking process. Appendix C provides the format for reporting a UXO hazard; Appendix E is a tool for plotting and tracking UXO hazards; and Appendix F is useful for tracking UXO and obstacle hazards and understanding the engineer obstacle overlay. Every unit is responsible for tracking UXO hazards in their battlespace. Units obtain this information from individual spot reports, situational awareness, and the engineer obstacle overlay prepared by the joint force engineer. The primary means of disseminating UXO information is through the engineer obstacle overlay.

b. **Tracking Prioritization.** Until each UXO hazard is cleared, the joint force engineer tracks all known UXO hazards, with assistance from the joint force EOD staff officer. The joint force engineer tracks, on the obstacle overlay, all known UXO that may operationally impact friendly forces. If excessive UXO is likely, the JFC may direct the engineer to prioritize the tracking effort. The JFC bases the tracking level on the location, amount, and threat of potential UXO hazards. In these cases, the EOD and engineer staff officers must prioritize tracking the hazards, presenting the greatest danger and those most operationally significant to the force. The joint force EOD officer, in coordination with the joint force engineer, maintains records and other historical information on reported UXO hazard areas. This information is used by follow-on units and units conducting civil-military or post-conflict operations. Regardless of the density of the UXO hazard, the joint force engineer will always track all minefields.

Chapter III
Service Component Capabilities

1. Background

This chapter outlines individual service EOD and engineer missions, C2 structures, and operational planning considerations in support of UXO operations. This chapter implements the following standardization agreements (STANAGs):

a. STANAG 2079, *Rear Area Security and Rear Area Damage Control.*

b. STANAG 2143, *Explosive Ordnance Reconnaissance/Explosive Ordnance Disposal.*

c. STANAG 2377, *Procedures for the Management of an EOD Incident for Use When Working With Other Agencies.*

d. STANAG 2389, *Minimum Standards of Proficiency for Trained Explosive Ordnance Disposal Personnel.*

e. STANAG 2485, *Engineer Countermine Operations.*

f. STANAG 2834, *The Operation of the EOD Technical Information Center.*

g. STANAG 2929, *Airfield Damage Repair.*

h. STANAG 2957, *International System used by Armed Forces in the Nuclear Field.*

2. U.S. Army

a. **EOD.**

(1) *Mission.* The U.S. Army EOD mission is to support the JFC by providing the capability to neutralize hazards from conventional UXO, NBC, and associated materials, and IED (both explosive and NBC), that present a threat to operations, installations, personnel and/or materiel. Army EOD forces can also dispose of hazardous foreign or U.S. ammunition, UXO, individual mines, booby-trapped mines, and chemical mines. EOD provides the Army with a rapidly deployable support package for the elimination of hazards from UXO in any operational environment. The EOD force neutralizes UXO that is restricting freedom of movement and denying access to supplies, facilities, and other critical assets. Army EOD forces equip, train, and organize to support tactical land forces across the spectrum of operations.

(2) *C2.* The EOD group provides C2 for all Army EOD assets and operations in theater. When directed by the JFC, the EOD group may become the nucleus of a subordinate EOD-focused JTF. The EOD battalions provide C2, mission tasking, technical intelligence acquisition/management, and limited administrative and

logistics support for up to 10 EOD companies. The EOD group, battalion, and company commanders have the additional duty of performing as the EOD special staff officer to their respective service component or functional command. In the absence of a fully deployed EOD group, battalion, or battalion (-), the command element may deploy as the C2 element for Army EOD assets in a given operation. Refer to Army FM 4-30.12 (FM 9-15)) and FM 4-30.16/MCRP 3-17.2C/NTTP 3-02.5/AFTTP(I) 3-2.32 for additional guidance on Army EOD operations.

(3) *(3) Operational Planning.* The EOD group and EOD battalion provide staff planning for Army EOD operations throughout the Army Forces (ARFOR) AO, and in most operational situations in other land areas within the JOA as determined by the JFC. Thus, the EOD group commander has the additional duty of performing as the EOD special staff officer to the ARFOR commander. The EOD battalion commander would also serve as the EOD special staff officer at an Army based JTF and, possibly, at the coalition JTF level. In the absence of a deployed EOD group or EOD battalion, the senior ranking Army EOD officer also serves as the ARFOR EOD staff officer. The Army EOD commander is responsible for providing the EOD annex to all OPLANs/OPORDs. This ensures that EOD forces fully understand and support the ARFOR commander's operations and also provides for force protection throughout the AO.

b. **Combat Engineers.**

(1) *Mission.* The five primary engineer functions are mobility, countermobility, survivability, general engineering, and topographic engineering. The specific engineering missions concerning UXO are breaching, clearing, and proofing minefields. In extreme high-operational tempo or high-intensity combat missions, U.S. Army engineer or other non-EOD units may conduct limited reduction or clearing of non-mine UXO hazards, under the technical guidance of Army EOD forces. During the post-conflict phase, engineers also assist EOD forces in battlefield UXO clean-up operations, as required. JP 3-34, JP 4-04, FM 3-34 (FM 5-100), and FM 3-34.211 (FM 5-116), provide more details on specific engineer units and tasks.

(2) *C2.* Command authority over engineer units is given to a maneuver commander when he needs responsive engineer support. The command relationship is either attachment, OPCON, TACON, or operational command (OPCOM). Supporting relationships include direct support or general support. Direct support authorization occurs when the maneuver commander requires immediate responsiveness from engineers, but does not need task organization authority. General support authorization occurs when the higher headquarters requires central control and flexibility when employing limited engineer assets. Army C2 units in theater include—

- Engineer command, a major subordinate command of the Army service component command.

- Theater Army engineer brigade(s), the principal subordinate unit of the engineer command.

- Corps engineer brigade(s).
- Engineer group(s).
- Division engineer brigade(s) with subordinate battalions, companies, and platoons. The engineer platoon is the lowest level at which an engineer unit can effectively accomplish independent tasks. The maneuver commander gets optimum support when the engineer battalion, company, or platoon is under his direct control.

Note. Refer to FM 3-34 (FM 5-100) for additional information on engineer C2.

(3) *Operational Planning.* The primary Army engineer elements that conduct landmine operations are corps and division combat engineers. Each maneuver brigade in an armor or mechanized infantry division is normally task organized with a mechanized combat engineer battalion, with one engineer company supporting each battalion task force. Light infantry, airborne, or air assault divisions have a single organic battalion of combat engineers, with one company normally task organized to each maneuver brigade. Initial/interim brigades are also supported by an organic combat engineer company. It is common to task organize additional corps-level combat engineer units to divisions and brigades to weight the main effort. Adjustment of the level of engineer support occurs after a thorough mission, enemy, terrain and weather, troops and support available, and time available (METT-T) analysis. (The Army adds a letter "C" to the end of the acronym for civilian considerations [METT-TC].) Additional corps combat battalions operate in the division on an area or task basis. Corps combat engineer battalions reinforce division engineers and provide support throughout the corps area. The major engineer commitment is to keep the lines of communications and tactical march routes open to sustain committed forces, support the shifting of forces, and supporting corps or task force-directed attacks.

3. U.S. Marine Corps

a. EOD.

(1) *Mission.* The USMC EOD mission is to provide force protection in support of the Marine air-ground task force (MAGTF). The USMC EOD force neutralizes hazards from foreign and domestic (conventional and NBC) UXO and IEDs that present a threat to operations, personnel, or materiel. Additionally, Marine Corps EOD units provide technical intelligence on disassembly of ordnance and rendering munitions inert.

(2) *C2.* The force service support groups (FSSGs) provide support to Marine ground combat elements and combat service support elements, and the Marine wing support group provides support to Marine air combat elements. The EOD officer/team leader directs and coordinates the execution of EOD tasks in support of the commander's intent. The EOD teams within the MAGTFs coordinate with each other to ensure complete support of all units. The senior EOD officer has staff responsibility for all EOD-related matters during a contingency or conflict.

(3) *Operational Planning.* For planning, the Pacific/Atlantic Marine expeditionary force (MEF) logistics staff officer (G-4) manages EOD operations. The

senior EOD Marine for a future mission conducts all operational planning. This Marine is responsible to the commander for providing the EOD annex for OPLANs/OPORDs to ensure full EOD support in all phases of the operation.

b. **Engineers.**

(1) *Mission.* Marine Corps engineers increase the combat effectiveness of the Marine division by rendering close combat engineer support and by providing both tactical and service support. Engineers also increase the effectiveness of the landing force by accomplishing general engineering missions of a deliberate nature. Marine engineer tasks related to UXO include—

(a) Mobility, countermobility, and general engineering.

(b) Engineer reconnaissance.

(c) Minefield breaching and proofing operations.

(d) Construction of barriers for UXO effects mitigation.

(2) *C2.* The Marine Corps task organizes engineers to MAGTFs. Each element, except the command element, contains organic engineer support. The engineers enhance the mobility, countermobility, and survivability of the Marine division and provide close combat engineer support and limited general engineer support.

(3) *Operational Planning.* Engineers are organic to each MEF (the largest type of MAGTF) and include a combat engineer battalion in the division, engineer support battalion in the FSSG, an engineer section in each Marine wing support squadron of the Marine aircraft wing, and elements from a Naval construction regiment. These units are organic to the MEF. In general, task organization of MAGTFs occurs for a given mission and can vary in size so the actual level of engineer support will depend on the situation and mission.

4. U.S. Navy

a. **EOD.**

(1) *Mission.* The U.S. Navy EOD mission supports national security strategy by providing forces capable of conducting land and underwater detection, identification, render safe, recovery, field evaluation, and disposal of explosive ordnance.

(2) *C2.* The Navy organizes EOD forces to support the JFC. The JFC has OPCON of EOD forces through the fleet and numbered fleet commanders. The EOD commanders and staff within the EOD organizational structure provide mission prioritization, tasking and monitoring, limited logistics support, and intelligence collection and dissemination.

(3) *Operational Planning.* The Navy generally categorizes EOD operations into three types: maritime operations, contingency operations, and ordnance intelligence and acquisition.

(a) Maritime Operations. The Navy conducts EOD operations to enhance ship survivability, preserve fleet warfighting capabilities, and enable naval, expeditionary and joint forces to achieve and maintain battlespace dominance through the reduction or elimination of hazardous UXO threats. The Navy assigns the following EOD forces to aircraft carrier battlegroups for operational commanders to employ as necessary to meet theater objectives:

- Amphibious ready groups.
- Mine countermeasures groups.
- Special contingency operations at sea and ashore.
- Shore installations where continuing EOD requirements exist.

(b) Contingency Operations. When threats involve conventional ordnance or weapons of mass destruction, the Navy EOD force provides a flexible and interoperable partnership with Naval Special Warfare, Army, Marine Corps, and special operations forces.

(c) Ordnance Intelligence and Acquisition. Navy EOD personnel can recover ordnance items on land or underwater, make the ordnance explosively safe, and return the item for exploitation. EOD detachments gather immediate preliminary intelligence on threat ordnance in the field and disseminate the intelligence to those requiring it in the AO. Then, when possible, the detachment or other asset conducts a more detailed exploitation of the AO. Data collected contributes to the detection of enemy stockpiles and identification of launch platforms and tactics, and supports the development of render safe procedures and countermeasures.

b. **Engineers.**

(1) *Mission.* Naval construction force (NCF) or Seabee units provide the Navy, Marine Corps, geographic combatant commanders, and other agencies with rapid contingency response for base facility construction, battle damage repair, disaster recovery support, and humanitarian assistance. Seabee units are expeditionary, rapidly deployable, self-sustaining, and capable of defensive combat operations. When supporting the Marine Corps, Seabees reinforce and augment the limited general engineering and construction capabilities of the MAGTF to enhance and sustain MAGTF operations ashore. Underwater construction teams are NCF units specialized to support amphibious landings and provide logistics over-the-shore. NCF units have no EOD/UXO capability; however, with their horizontal and vertical construction capabilities, NCF units can support other units undertaking EOD/UXO missions. The type of support may include—

(a) Mobility improvements (bridge and road construction and repair).
(b) Construction of base camps.

(c) Rapid runway repair.
(d) Well drilling.

Note. Additional information on NCF unit capabilities is available in JP 3-34, JP 4-04, and NWP 4-04.1/MCWP 4-11.5.

(2) *C2.* The Navy organizes NCF units for quick and effective response as dictated by planning assumptions and guidance.

(3) *Operational Planning.* The NCF includes both Reserve units and active units with Reserve components. The Navy organizes one of its two amphibious construction battalions (ACBs) under each Naval beach group (NBG) of the Atlantic and Pacific Fleets. The fleet commanders exercise administrative control (ADCON) and OPCON of the NCF units through the Naval construction brigades and NBGs. For more detailed information on NCF units, refer to OPNAVINST 5450.46K and NWP 4-04.1/MCWP 4-11.5. For more detailed information on ACBs, refer to OPNAVINST 3501.93D.

5. U.S. Air Force

a. **EOD.**

(1) *Mission.* USAF EOD personnel protect people, resources, and the environment from hazardous ordnance caused by either accidents or enemy attacks. USAF EOD personnel also furnish services to render safe, dispose, and clear UXO affecting operations on fixed installations, dispersal bases, or in the assigned AO.

(2) *C2.* The USAF normally assigns EOD units to the base civil engineer supporting a USAF wing. The USAF task organizes EOD units as expeditionary civil engineer squadrons, EOD flights, or EOD elements. If no base civil engineer is assigned, the Air Force assigns EOD units to the senior USAF commander. During increased threat conditions, a USAF wing can establish a survival recovery center for wing C2. The senior EOD representative performs duties at the survival recovery center to control all EOD operations at the deployed location or regional support as required.

(3) *Operational Planning.* The USAF organizes EOD force packages into unit type codes (UTCs) to provide flexible structures to support contingency missions. Planners can combine the UTCs in building block fashion to provide coverage for location-specific missions. Refer to FM 4-30.16/MCRP 3-17.2C/NTTP 3-02.5/AFTTP(I) 3-2.32 for a comprehensive list of Air Force EOD UTCs and the *War and Mobilization Plan*, volume 1, Annex S, for instructions for building EOD unit capabilities using the various UTCs.

b. **Engineer.**

(1) *Mission.* USAF civil engineer (CE) units provide, operate, maintain, restore, and protect the installations, infrastructure, facilities, housing, and environment, as necessary to support air and space forces having global reach and power across the range of military operations.

(a) Prime Base Emergency Engineering Force (BEEF) is the primary organizational structure for supporting both mobility and in-place requirements. Prime BEEF's mobility mission is to go with a deploying flying squadron and bed down the flying unit's forward base of operations. Prime BEEF support to EOD forces includes—

- Force protection initiatives.
- Explosive ordnance reconnaissance.
- Augmentation for ordnance removal.
- Plotting, tracking, and identifying UXO.
- Heavy equipment support.
- Firefighting.
- NBC detection.

(b) Rapid Engineers Deployable Heavy Operational Repair Squadron, Engineers (RED HORSE) squadrons are theater assets with regional responsibilities and limited demolition capabilities. RED HORSE squadrons and their associated UTC configurations provide highly mobile, largely self-sufficient, rapidly deployable echelons to support major force bed down requirements and to repair heavy war damage. They are stand-alone units not tied to peacetime base support. They provide air component commanders a dedicated, flexible, airfield and heavy construction and repair capability. This allows the geographic combatant commander to move and support missions as the air order of battle dictates. RED HORSE squadrons provide a heavier engineering capability than Prime BEEF units. In their deployed mode, RED HORSE units are regional assets, not tied to a specific weapon system, and are not responsible for day-to-day operations and maintenance at any specific location. RED HORSE squadrons accomplish major construction work in forward locations. They provide bed-down support for the larger air expeditionary force deployments. Because of this, RED HORSE squadrons require an organic logistics and contracting capability to include vehicle maintenance, food service, supply, and logistics.

(2) *C2.*

(a) The USAF task organizes Prime BEEF mobility units that are task organized as expeditionary CE groups, expeditionary CE squadrons, or CE flights assigned to support specific operating locations. The senior officer assigned to the Prime BEEF unit acts as the battlefield coordination element.

(b) The Air Force task organizes RED HORSE units as RED HORSE groups, RED HORSE squadrons, or RED HORSE flights assigned to the theater USAF component commander as a theater asset.

(3) *Operational Planning.* The Air Force can tailor UTCs to provide specific engineer capabilities. The *War and Mobilization* Plan, volume 1, Annex S describes Prime BEEF and RED HORSE UTCs.

Appendix A
Identifying UXO

1. Purpose

This appendix implements STANAG 2143, edition 4. Recognizing a UXO hazard is the initial and most important step in reacting to a UXO hazard. All ordnance can incorporate anti-disturbance or anti-removal devices; therefore, clearing any UXO without proper training, tools, and information is extremely dangerous. Indiscriminate destruction of ordnance without positive identification of type, by function and filler, can result in extreme consequences. Friendly and threat forces use a multitude of ordnance manufactured in all shapes and sizes. This appendix explains and provides examples of the general identifying features of the different types of ordnance, of both U.S. and foreign manufacture. Generally, there are four main ordnance types (corresponding to line 4 of the UXO Spot Report): dropped, projected, thrown, and placed.

2. Identification References

Several DOD agencies publish guides for ordnance identification:

a. **DOD.** Ordnance Data Series. This is an excellent ordnance identification and safety compact disk series used throughout the DOD and by civilians.

b. **Army.** FM 4-30.11 (FM 21-16) and Technical Manual (TM) 43-series (see the Army reference section of this publication).

c. **Marine Corps.** The Marine Corps uses Army and Navy sources.

d. **Navy.** U.S. Naval Sea Systems Command, *U.S. Naval Mine Familiarization*, January, 2000.

e. **Air Force.** Air Force handbook (AFH) 32-4014 and Air Force visual aid (AFVA) 32-4022. (See http://afpubs.hq.af.mil.)

3. Dropped Ordnance

Regardless of its type or purpose, dropped ordnance is dispensed or dropped from an aircraft. There are three subgroups of dropped ordnance: bombs, dispensers (containing submunitions), and submunitions. Consider all bombs and submunitions to have magnetic/seismic or anti-disturbance fusing. Simply stated, this means approach could detonate the ordnance. Observe all ordnance from the safest distance (82 feet or 25 meters for most ordnance), with binoculars if necessary, that still allows the gathering of necessary information.

a. **Bombs.** General-purpose bombs of all countries are similar in construction to the examples at Figure A-1. About 50 percent of the weight of most bombs is explosive filler, and may contain high explosives, chemicals, or other hazardous materials. Observe all safety precautions, such as evacuating personnel, closing off areas, and preventing unnecessary activities near the unexploded bomb.

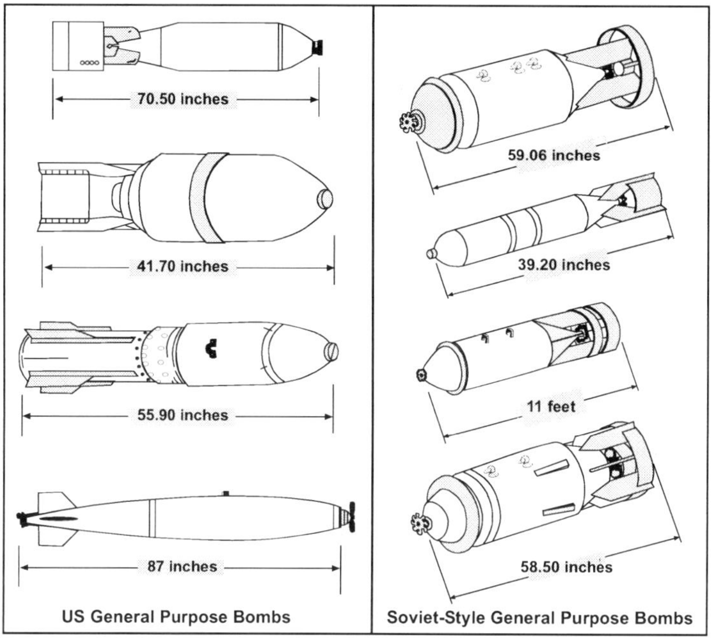

Figure A-1. U.S. and Soviet-Style General Purpose Bombs

b. **Dispensers.** Dispensers are another type of dropped ordnance. Like bombs, they are carried by aircraft. The dispenser payload, submunitions, is smaller ordnance. Figure A-2 provides examples of submunition dispensers and illustrates where the submunitions are dispersed from the dispenser body. Dispensers come in a variety of shapes and sizes, depending on the payload inside. Never approach a dispenser or any part of a dispenser, because the payload of submunitions scatters in the area where the dispenser hit the ground.

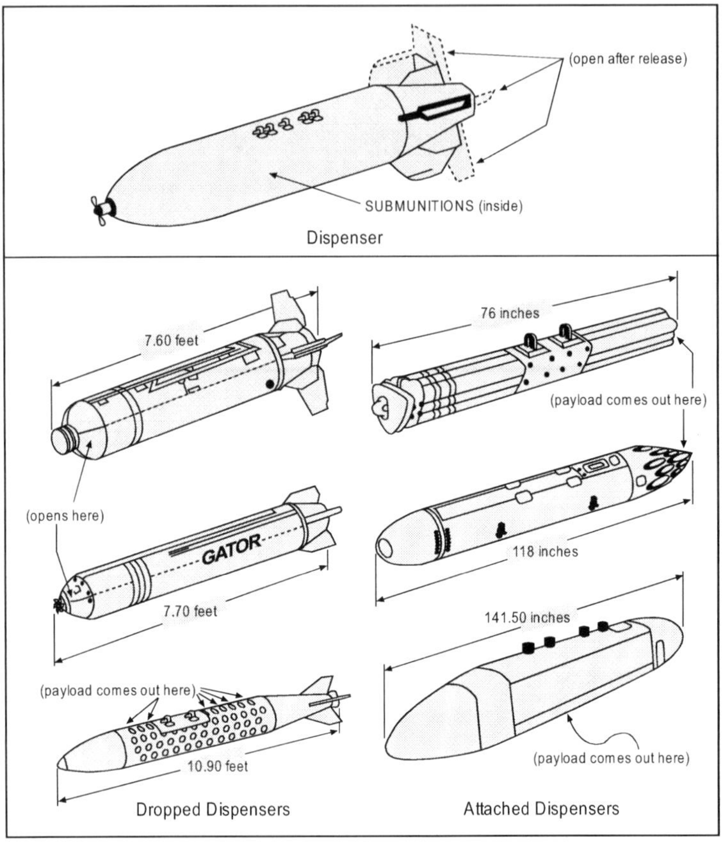

Figure A-2. Dropped and Attached Dispensers

c. **Submunitions.** Classifications of submunitions are either bomblets, grenades, or mines. They are small, explosive- or chemical-filled, items designed for saturation coverage of a large area. See Figure A-3 for examples of submunitions.

(1) *Bomblets and Grenades.* Dispensers, missiles, rockets, or projectiles spread submunitions; these may be either antipersonnel (APERS), anti-materiel (AMAT), antitank (AT), dual-purpose (DP), incendiary, or chemical. Each of these delivery systems disperses its payload of submunitions while still in flight; the submunitions drop over the target.

(2) *Mines.* This is a large category of submunitions. It is important to understand the difference between scatterable and placed mines for accurate reporting. The major difference between scatterable mines and placed mines is that

the scatterable mines land on the surface and are often more easily visually detected. Placed mines, discussed later in paragraph 6 of this appendix, may be placed on the surface, hidden, or buried under the ground.

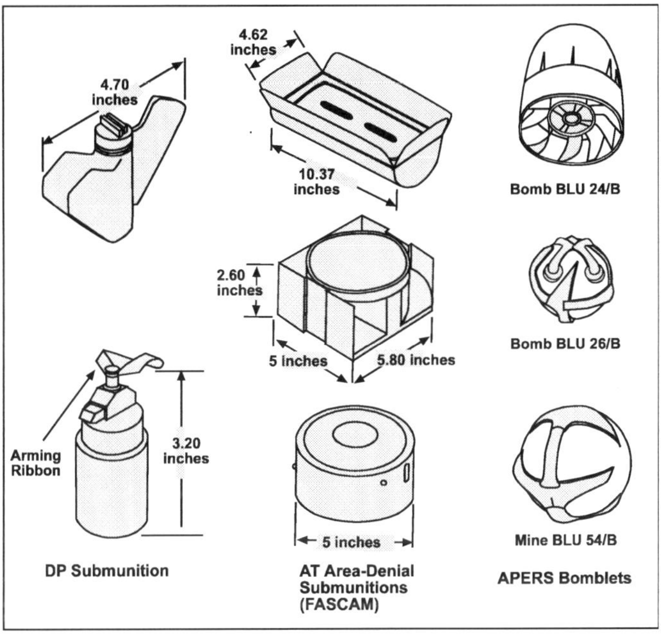

Figure A-3. Submunition Examples

4. Projected Ordnance

Projected ordnance can be either projectiles, mortars, rockets, rifle grenades, or guided missiles.

a. **Projectiles, Mortars, Rockets, and Rifle Grenades.** All projected ordnance is fired from some type of launcher or gun tube. Projected ordnance falls into the following five subgroups: projectiles, mortars, rockets, rifle grenades, and guided missiles. The next paragraph discusses guided missiles. Projected ordnance is either solid metal, filled, or partially filled with an explosive, chemical warhead, or submunitions. These items may not detonate on impact. Personnel must not attempt to move or disturb these items. Like bombs, these items can have impact or proximity fusing. They can also be fused with time-delay fusing that functions at a preset time after firing. Getting too close to proximity fusing will cause the fuse to function and the UXO will explode. Examples of projectiles, mortars, rockets, and grenades are shown at Figure A-4.

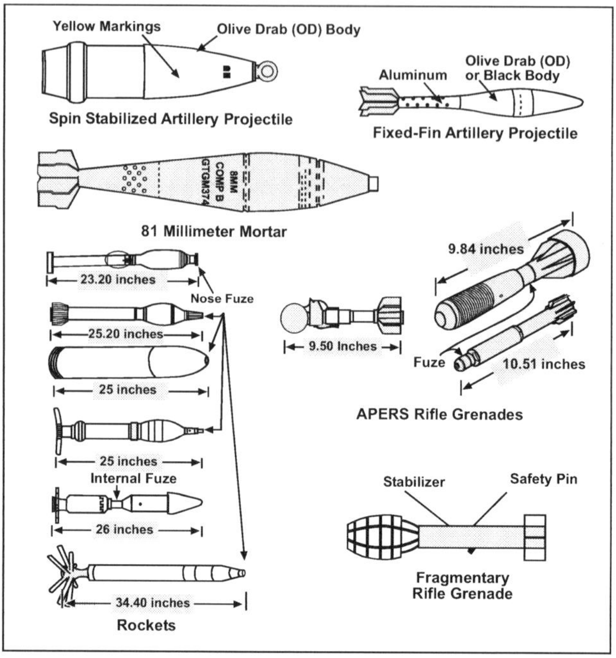

Figure A-4. Projectiles

b. **Guided Missiles.** Guided missiles allow for control of the munitions after firing, and are guided to their target by various guidance systems. The tube-launched, optically tracked, wire-guided (TOW) missile is an example of a firer-guided missile. (See Figure A-5.) Guided missiles have internal, proximity fusing. Therefore, personnel should not approach any guided missile.

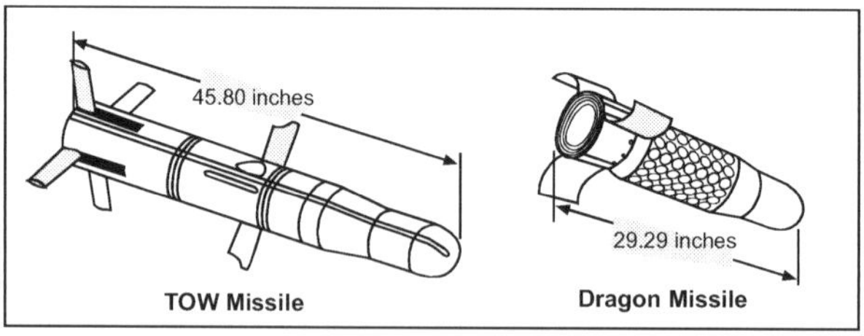

Figure A-5. Guided Missiles

5. Thrown Ordnance

Commonly known as hand grenades, classification of thrown ordnance is by use as follows: fragmentation (also called defensive), antitank, smoke, and illumination. Moving, jarring or otherwise disturbing this ordnance may cause it to explode. Never pick up or disturb a hand grenade, even if the spoon and safety pin are still attached. Consider all grenades to incorporate anti-disturbance or anti-removal devices. (See Figure A-6.)

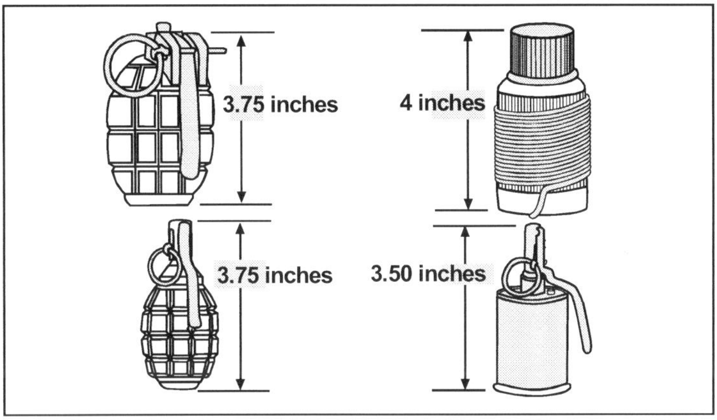

Figure A-6. Thrown Ordnance (Fragmentary Grenades)

6. Placed Ordnance

Placed ordnance is commonly referred to as land mines. Land mines are hidden, buried, or placed on the surface, and often cannot be seen. Visual detection of land mines is often difficult. Consider all mines to have anti-disturbance or anti-removal devices. Mines equipped with magnetic or seismic influence fuses may detonate when disturbed. Conduct all observation of this ordnance, with binoculars, at the greatest distance that still allows gathering of required information. Placed land mines destroy vehicles and inflict casualties on personnel who step on or drive across them. There are three basic types of land mines: AT, APERS, and chemical. (See Figure A-7.)

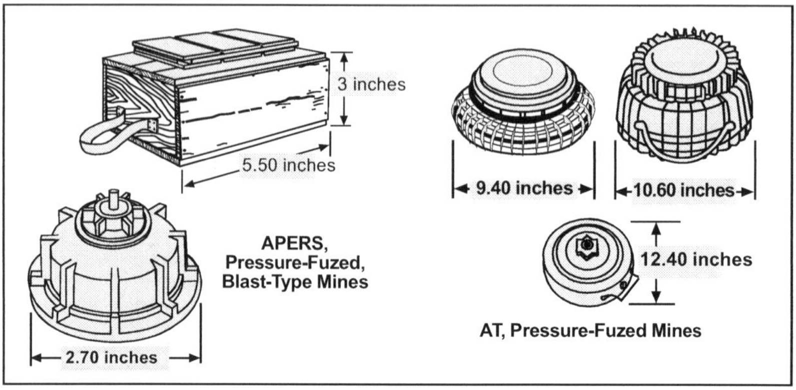

Figure A-7. Placed Ordnance

7. Sea Mines

Sea mines, also called naval mines, are those mines emplaced in deep or shallow waters, coastal areas, harbor entrances, rivers, canals, and estuaries. Aircraft, submarines, or surface ships emplace sea mines. Sea mines may also be hand-laid. Observe all safety precautions associated with ordnance having influence or anti-disturbance fusing by evacuating personnel and equipment, securing access to the area, and preventing unnecessary activities near the mine. The two major categories of sea mines are antisubmarine and surface ship mines (see Figure A-8), and very shallow water (VSW) anti-landing or amphibious mines (see Figure A-9).

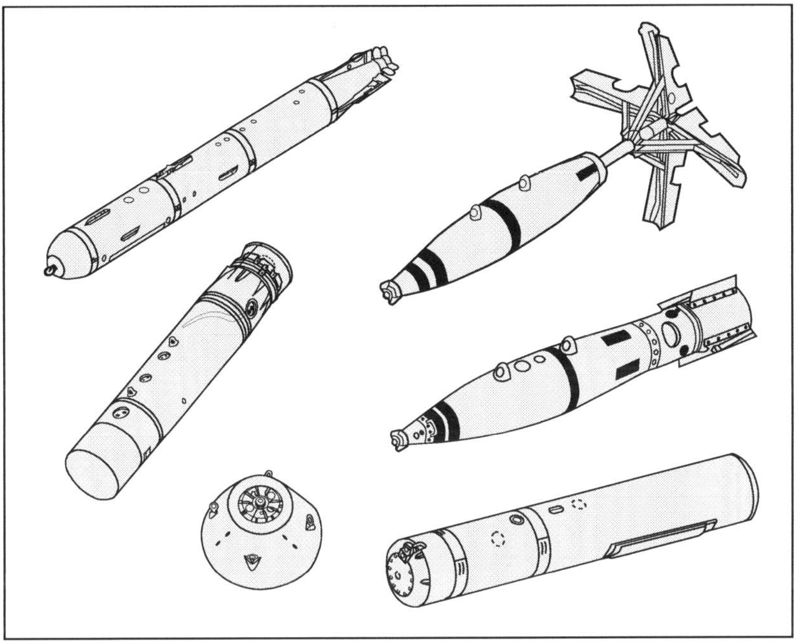

Figure A-8. Antisubmarine and Surface Ship Mines

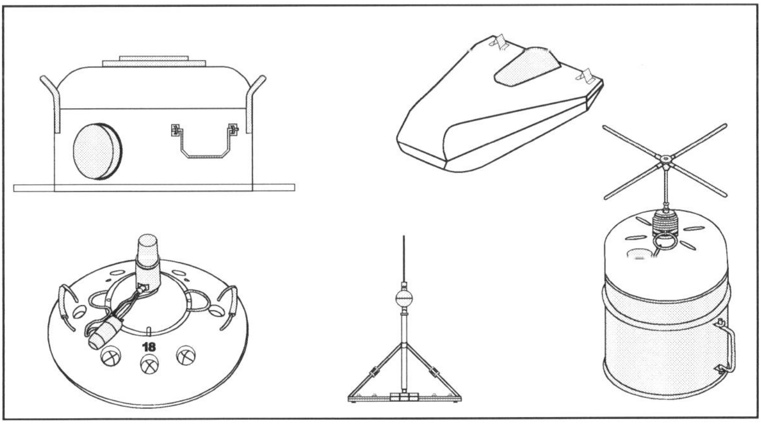

Figure A-9. Very Shallow Water/Anti-Landing/Amphibious Mines

Appendix B
Marking UXO

1. Confirm the Presence of UXO

This appendix implements STANAG 2002. The strict observance of basic safety precautions and sound judgment lessen the danger of UXO.

a. **Safe Distance.** Upon identification of the UXO, retire to a safe distance and enforce evacuation measures, when evacuation is possible. Do not remain in the immediate danger area any longer than absolutely necessary. See Appendix D, Table D-1 for evacuation distances.

b. **Disturbances.** ***Do not touch or disturb*** the UXO or associated components, including loose wires or parachutes. Disturbances, either mechanical or otherwise, may cause the item to detonate.

c. **Chemical/Biological Hazards.** Assume the presence of chemical agents if you detect the presence of liquid droplets, dead animals, dissolved paint, or peculiar odors. Put on your protective equipment immediately.

d. **Radio Transmit Hazard.** ***Do not transmit or key radios*** within 100 meters of a UXO. It may cause a detonation.

2. Mark the UXO

a. **Marker Description.** Marking a mine or UXO hazard is just as important as marking other hazard areas, such as NBC contaminated areas. When marking a UXO or mine hazard, use the standard triangular signs, if available, that indicate danger by their background color of red. Order the standard mine marker using NSN 6230-00-926-4336, and order the standard UXO marker using NSN 7690-01-463-3422. Both are available through service supply systems, and are shown in Figure B-1. The background is red with a white inset indicating the type of UXO.

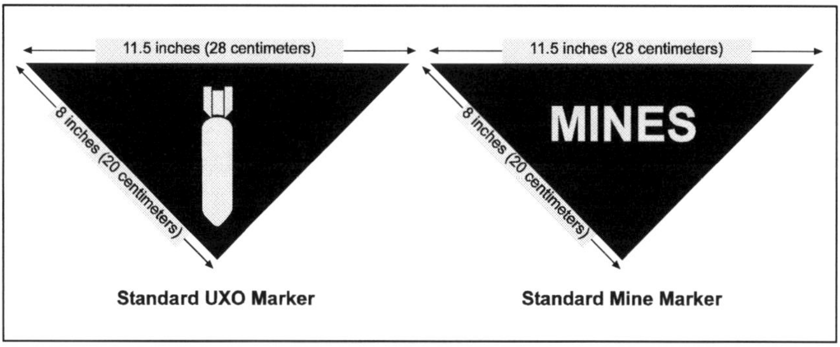

Figure B-1. Standard UXO and Mine Markers

b. **Marker Placement.** Place the markers above the ground at waist level (about 3 feet) with the point down as shown in Figure B-2. Place the marker no closer to the hazard than the point at which you first recognize the hazard. Attach the marker to a stake (as shown in Figure B-2), a tree, or other suitable holder. Make sure that the marker is clearly visible from all directions.

Figure B-2. Elevated Marking of UXO/Mines

c. **Marker Visibility.** Mark all logical approach routes to the area. If the hazard is near a road, put a marker on each side, as a minimum, of the road approaching it.

d. **Field Expedient Methods.** If standard markers are not available, use other suitable materials (such as engineer tape or colored ribbons). Use the same color marker and material to avoid confusion. When using field-expedient materials, the same marking principles apply. That is, place about three feet off the ground and clearly visible from all approach routes. See Figure B-3.

Figure B-3. Field Expedient Marking Method

e. **Minefield or UXO Area Hazard Marking Considerations.** See Figure B-4 for an example of marking a minefield or other UXO area hazard. This example uses the mine marker. Marking of UXO uses a different marker, but all other procedures for marking mines and UXO are identical. If there is a large concentration of hazards such as submunitions, mark the area as you would a scatterable minefield by placing at least one sign every 15 meters around the area. As a general rule, the hazard itself must be easily seen from any of the markers.

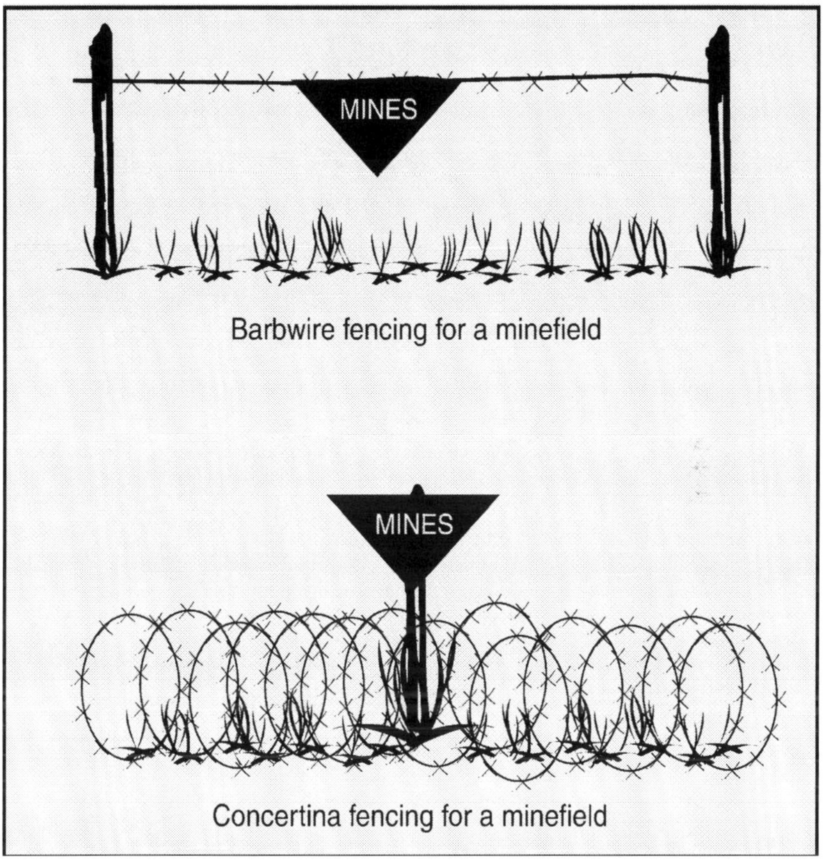

Figure B-4. Examples of Marking a Minefield or UXO Area Hazard

Appendix C
Reporting UXO

1. Purpose

This appendix implements STANAG 2143. The UXO Spot Report is a timely and detailed, two-way reporting system; it clearly identifies the location of the UXO hazard, briefly describes the hazard, the priority for clearing, and the unit(s) affected. The UXO Spot Report is the first-echelon report that is sent when an observer detects UXO. Units can use the report to request help in handling a UXO hazard if the unit cannot reduce the hazard either by using protective works or moving away from the UXO. The report also assists the commander to set priorities based on the situation. The report consists of nine lines and is sent by the fastest means available. See Table C-1.

2. Routing

Forward the UXO Spot Report through the chain of command. Each commander in the chain who receives/reviews the report may change the priority to reflect the current tactical situation or projected battle plan. Each commander in the chain is responsible for forwarding UXO Spot Reports through command channels and ensuring that the proper priority is set for each report. When a commander changes the original priority, the higher-level commander must inform subordinate commands, especially the initial reporting unit. Based on METT-T (with the Army adding a "C" for civil considerations), engineer forces use the UXO Spot Reports to prioritize and sequence their response to breaching, clearing, and proofing mobility obstacles. EOD forces use the UXO Spot Report to help sequence their response to all UXO incidents.

Table C-1. UXO Spot Report

Line 1.	Date-Time Group (DTG) the item was discovered.
Line 2.	Reporting Activity (unit identification code) and location (grid of UXO).
Line 3.	Contact Method: Radio frequency, call sign, point of contact, and telephone number.
Line 4.	Type of Ordnance: Dropped, projected, placed, or thrown. If known, give the size of the hazard area and number of items. Without touching, disturbing, or approaching (tripwire) the item, include details about size, shape, color and condition (intact or leaking).
Line 5.	NBC Contamination: If present, be as specific as possible.
Line 6.	Resources Threatened: Report any threatened equipment, facilities, or other assets.
Line 7.	Impact on Mission: Provide a short description of your current tactical situation and how the presence of the UXO affects your status.
Line 8.	Protective Measures: Describe any measures taken to protect personnel and equipment.
Line 9.	Recommended Priority: Recommend a priority for response by EOD technicians or engineers.

Priority	Basis
Immediate	Stops the unit's maneuver and mission capability or threatens critical assets vital to the mission.
Indirect	Slows the unit's maneuver and mission capability or threatens critical assets important to the mission.
Minor	Reduces the unit's maneuver and mission capability or threatens non-critical assets of value.
No Threat	Has little or no affect on the unit's capabilities or assets.

Note. This report does not supersede, unless specified, specific service reporting requirements.

Appendix D
Protective Measures

There are three methods to protect personnel and equipment: evacuate, isolate, or barricade (or any combination thereof).

1. Evacuate

When possible, evacuation of all personnel and equipment is the best protective measure. See Table D-1 for minimum evacuation distances. Allow occupation of the area only by mission essential personnel and equipment. After evacuating all personnel and equipment, allow movement within the area only for essential operations. Use barricades to protect mission-essential personnel and equipment that must stay in the area. (See paragraph 3, this appendix for additional information on barricades). Personnel who must remain in the area should also wear all protective equipment (to include Kevlar helmets and vests).

Table D-1. Evacuation Distances

Explosive Weight (pounds)	Evacuation Distance (meters)
27 or less	300
30	310
35	330
40	350
45	360
50	375
100	475
150	550
200	600
250	625
300	675
400	725
500	800

2. Isolate

Sometimes, for mission-related, operational, or other reasons, evacuation of personnel or equipment is not possible or it is not possible to leave a particular area. In these situations, isolate the assets (personnel, equipment, and operations) from the UXO by establishing a safe area and limiting exposure.

3. Barricade

A barricade provides limited protection by blocking blast and fragmentation from an explosion. "Suppressive" barricades are constructed to isolate an explosion, to deflect the thermal/shockwave, and absorb low-angle, high-speed fragments. "Protective" barricades are constructed around exposed resources to shelter from overpressure and impact of high-angle, low-speed fragments, which can escape over the top of "suppressive" barricades. Natural terrain features can provide adequate frontal and overhead protection. Qualified engineering, EOD, or safety personnel should assess the potential of existing natural terrain barricades. If natural barricades are judged inadequate, construct artificial barriers. Constructing barricades is very time consuming and requires a large number of sandbags or earth-moving equipment. While building barricades, personnel should not disturb the UXO and should wear all protective equipment (to include Kevlar helmets and vests). Depending on the size of the UXO, build suppressive barricades around the UXO to protect the entire area, or build protective barricades next to non-evacuated personnel or equipment. Priority should be given to evacuation of resources and building protective barricading. See Figure D-1.

Figure D-1. Barricade Examples

a. **Small Yield UXO.** For small yield UXO, generally less than three inches in diameter, a double-wall thickness of sandbags should surround the area of the UXO. Stack the sandbags to at least three feet high and thick enough (minimum two sandbags deep) to protect personnel and equipment from the blast and fragmentation.

b. **Medium Yield UXO.** For medium yield UXO, generally from three inches to seven inches in diameter, a wall thickness of four or five sandbags should surround the area. Stack the sandbags to a height of at least five feet to protect assets.

c. **Large Yield UXO.** UXO over seven inches in diameter are generally too large to build effective barricades around them. In these cases, barricade the equipment and personnel activity areas. Marines and soldiers can use FM 21-16/MCRP 3-17.2A for additional guidance on protective measures.

Appendix E
UXO Graphics

1. Purpose

This appendix discusses procedures and provides examples of map symbols used with map overlays for tracking/plotting UXO. All units can benefit by tracking and maintaining situational awareness of UXO within their AO. Operational and tactical commanders and staffs can utilize the graphics in this appendix to track all UXO hazards within their AO.

2. Procedures

a. **Plotting Requirements.** Every unit should maintain locations of UXO within their battlespace. EOD and engineer units are responsible for accurately recording the location of UXO incidents within their AO. This includes, at a minimum, DTG reported, type/quantity, grid location, and DTG the UXO was cleared. As additional information becomes available, the joint force engineer disseminates the information through the J-2/J-3 to other organizations operating within the JOA. The joint force EOD officer will provide appropriate information to the joint force engineer to incorporate into the engineer obstacle overlay.

b. **Plotting Procedures.** Plot UXO on an overlay or on a map using the following guidelines:

(1) *Graphic Symbols.* Three types of graphic symbols may be used when plotting UXO: point UXO, area UXO, and graphics for mines/minefields.

(a) Point UXO. The graphic control measure to depict a point UXO incident is an inverted triangle with the letters "UXO" printed in the center of the triangle. The bottom point of the triangle represents the exact location of the UXO; using the extended line option, the open end of the extended line represents the exact location of the UXO. The joint force EOD officer may assign a numerical value to the hazard for consistent tracking and historical information. See Figure E-1.

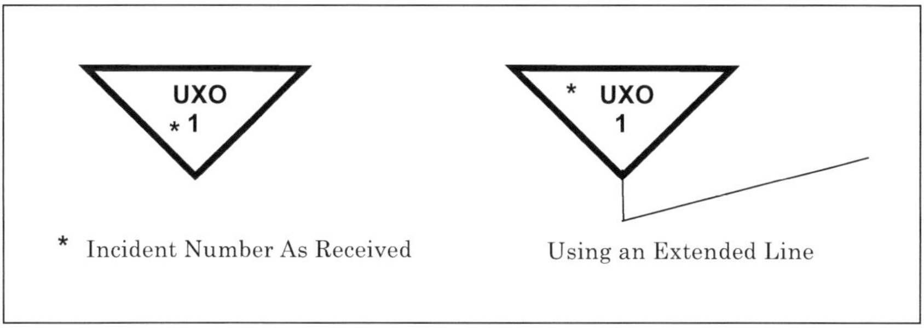

Figure E-1. Graphic for a Point UXO

(b) UXO Area. The graphic symbol for a UXO area is a line drawn around the perimeter of the area, with "UXO" and a numerical value relating to its order of reception placed on both sides of the perimeter line. The area in the center of the UXO hazard area will include the DTG of discovery or when the incident was fired, whichever is earlier, the type of UXO (including either U.S. or foreign), and the DTG confirmed. See Figure E-2.

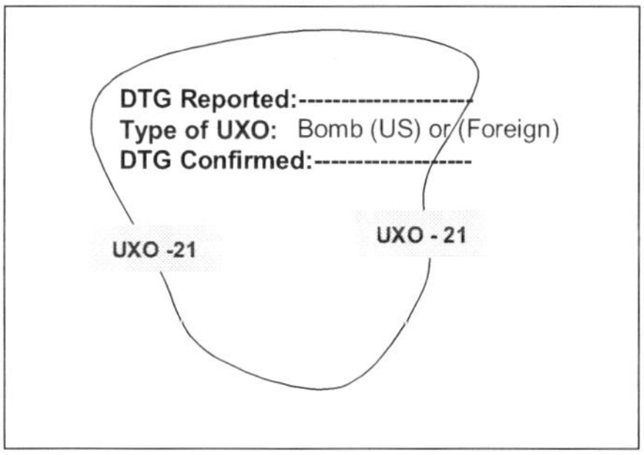

Figure E-2. Graphic for a UXO Area

(c) Mine/Minefields. There are several different graphic symbols for mines/minefields to provide a better visual graphic of the hazard. Figure E-3 provides a comprehensive list of available graphics for mines/minefields.

(2) *Abbreviations.* When using abbreviations to describe the type of UXO, use the abbreviations in Appendix F, Table F-2 to refer to the specific UXO type.

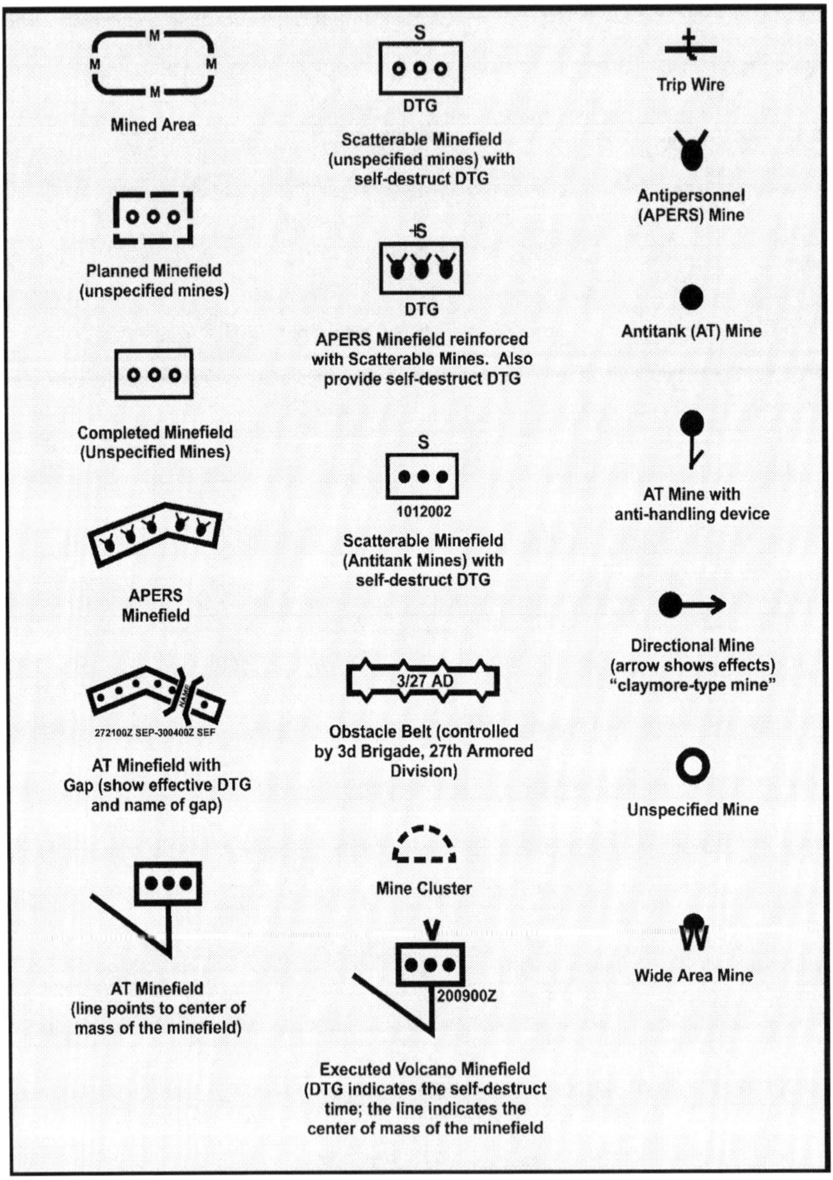

Figure E-3. Mine and Minefield Graphics

(3) *Colors.* UXO are obstacles, and all obstacles (friendly, enemy, neutral, or factional) will be drawn using the color green. If the need to track enemy-generated UXO arises, use the abbreviation "ENY" placed on the graphic in at least two places.

(4) *Legends.* A legend on the overlay assists the user to understand the information presented on the overlay. The legend is a table at the bottom of the overlay that reflects: the appointed UXO number, type of hazard, location, DTG discovered/fired, and the DTG confirmed or cleared (if applicable). See Figure E-4 for an example of a UXO overlay and legend. This overlay is useful to non-EOD or engineer units that do not desire to track UXO to the same level of detail as an engineer obstacle overlay.

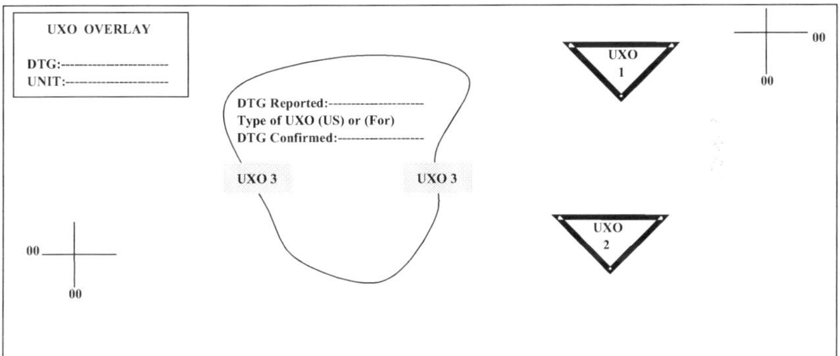

Number	Type	Grid Location	DTG Discovered/Fired	DTG Cleared
1	Bomb (6 each)	PV 12784563	120730S MAR00	Complete 12 1530S MAR00
2	Projectile (1 each)	PV 23375638	121045S MAR00	
3	DPICM (3 each)	PV 45341278	131630S MAR00	

Figure E-4. UXO Overlay and Legend Example

Appendix F
UXO and Obstacle Numbering

1. Purpose

This appendix describes the numbering system engineer units use on engineer obstacle overlays to label UXO and obstacles. Both the Army and Marine Corps use the same numbering system. This appendix uses Army units for example only. All units tracking locations of UXO hazards and obstacles can benefit by understanding the numbering system because of the information it provides. The numbering system allows the obstacle overlay user to identify the unit responsible for the obstacle, the type of obstacle or hazard, and its status. While all units do not need to maintain UXO/mine hazard status in the detail used by an engineer/EOD unit, this appendix can be used as a reference to better understand the engineer obstacle overlay as questions arise.

2. UXO and Obstacle Numbering

a. Introduction. The UXO/obstacle numbering system applies to friendly and enemy UXO/obstacles, but there is a significant difference when numbering friendly obstacles. Numbering friendly, emplaced obstacles uses 11 alphanumeric characters, and a 12th character to provide the obstacle status. Enemy-emplaced obstacles and hazards use only four characters, and a fifth character to provide the obstacle/hazard status. Engineers use all 12 characters in the table for numbering friendly obstacles. Characters 1 through 7 do not apply to enemy-emplaced obstacles and hazards. When numbering enemy obstacles/hazards, use only characters 8 through 12. See Table F-1.

Table F-1. UXO/Obstacle Numbering System

CHARACTER	DESCRIPTION
1 through 4 (friendly emplaced only)	Alphanumeric descriptions of the headquarters type and numerical designation that established the obstacle zone. Character 1 designates the unit type with a letter (A, armored division; C, Cavalry division; I, Infantry division; and Z, Corps); character 2 through 4 designate the actual unit.
5 (friendly emplaced only)	Letter indicating the obstacle zone. If there is no obstacle zone, an asterisk is used.
6 (friendly emplaced only)	Number indicating the belt number in the obstacle zone. If there is no belt, an asterisk is used.
7 (friendly emplaced only)	Letter indicating the group in the obstacle belt. If there is no group associated with the belt, an asterisk is used.
8 and 9 (friendly or enemy)	Letters indicating the obstacle/hazard type. See Table F-2.
10 and 11 (friendly or enemy)	Two numbers indicating the obstacle/hazard number. Numbers are added in the order in which they are planned (friendly) or reported (enemy).
12 (friendly or enemy)	One of four characters indicating the obstacle status. If this character is blank, the obstacle has been cleared. / (slash) = planned obstacle - (dash) = obstacle being prepared + (plus) = prepared, but not executed obstacle. (The + is for reserve demolition targets and may indicate a readiness state of safe or armed). X = executed or completed obstacle.

b. **Characters 1 through 4: Unit Designation.** The first four characters of the numbering system uses a letter designation for the headquarters, followed by three numbers to designate the actual unit. The first character is normally provided in the OPLAN/OPORD within the Engineer Annex to identify the major headquarters. For example, XVII Corps is shown as Z017; 23rd Armored Division is shown as A023; and the 103d Airborne Division is shown as I103.

c. **Character 5: Obstacle Zone.** Obstacle zones are a graphic control measure that Corps and division commanders use to grant obstacle-emplacement authority to brigades or other subordinate units. Corps and division commanders also use zones to ensure that subordinates emplace obstacles that support the higher commander's scheme of maneuver and that do not interfere with future operations. To indicate an obstacle zone, use a single letter starting with "A" and continuing in sequence. The user can distinguish obstacle zone A in XVII Corps from obstacle zone A in 77th Infantry Division by the headquarters designation (Z017-A versus I077-A). The designation for the second obstacle zone in 77th Infantry Division is I077-B.

d. **Character 6: Obstacle Belts.** Obstacle belts are the graphic control measure that brigade commanders use to contain tactical obstacle employment. They plan obstacle belts within assigned obstacle zones to grant obstacle-emplacement authority to their major subordinate units. Obstacle belts focus obstacles in support of the brigade scheme of maneuver and ensure that obstacles do not interfere with the maneuver of any higher headquarters. Indicate obstacle belts using a single number following the obstacle zone designator, starting with "1".

e. **Character 7: Obstacle Groups.** Obstacle groups are one or more individual obstacles grouped to provide a specific obstacle effect. Task forces use obstacle groups to ensure that company teams emplace individual obstacles that support the task force scheme of maneuver. In rare cases, brigades, divisions, or even Corps may use obstacle groups for specific tactical obstacles. Also, units integrate obstacle groups with direct- and indirect-fire plans in detail. Depict obstacle groups with a single letter starting with "A," and continue in sequence. For example, the first obstacle group in obstacle belt 2 of obstacle zone B in the 77th Infantry Division is I077-B2A.

f. **Characters 8 and 9: Obstacle/Hazard Type.** There are numerous types of obstacles and hazards, and the use of characters 8 and 9 provide the type of obstacle/hazard at a particular site. Table F-2 provides a list of UXO hazards and obstacle types. Building on the previous example, an antitank ditch emplaced by the 77th Infantry Division is shown as I077-B2A-AD. A standard block minefield in the same scenario is shown as I077-B2A-MB. For a UXO hazard, the obstacle hazard type would represent the first and second character of the numbering system. A UXO hazard with CBUs is shown as UC, or an enemy antitank ditch is shown as ED.

Table F-2. UXO/Obstacle Type Abbreviations

Abbreviation	Description
Miscellaneous Obstacles	
AB	Rubble by blade
AD	Antitank ditch
AE	Rubble by explosives
AH	Log hurdles
AL	Log crib, log obstacles
AM	Moveable military operations on urbanized terrain (MOUT) obstacle (as an example, a car or bus)
AN	Expedient, nonstandard
AP	Post obstacles (as an example, a hedgehog or tetrahedron)
AR	Rubble by combat engineer vehicle gun
AT	Abatis

Table F-2. UXO/Obstacle Type Abbreviations

Abbreviation	Description
	Bridge Demolitions
BA	Bridge demolition, abutment
BC	Bridge demolition, combination of abutment and span
BS	Bridge demolition, span
	Enemy Ordnance Hazards
EB	Enemy bridge demolition
ED	Enemy antitank ditch
EE	Enemy rubble obstacle
EL	Enemy log obstacle
EM	Enemy minefield
ER	Enemy road crater
ES	Enemy scatterable minefield
ET	Enemy abatis
EW	Enemy wire obstacle
EX	Enemy unidentified hazard
	Minefields
MB	Minefield, block
MD	Minefield, disrupt
MF	Minefield, fix
MN	Minefield, nonstandard pattern
MP	Minefield, protective
MQ	Minefield, nuisance
MS	Minefield, standard pattern
MT	Minefield, turn
	Road Craters
RD	Road crater, deliberate
RH	Road crater, hasty
RM	Road crater, mined
	Family of Scatterable Mines (FASCAM)
SA	FASCAM, area-denial artillery munitions (ADAM)
SB	FASCAM, Gator
SF	FASCAM, ADAM, remote antiarmor mine (RAAM) system
SG	FASCAM, ground-emplaced, mine-scattering system (GEMSS)
SM	FASCAM, modular pack mine system (MOPMS)
SP	FASCAM, pursuit deterrent munitions
SR	FASCAM, RAAM
SV	FASCAM, Volcano

Table F-2. UXO/Obstacle Type Abbreviations

Abbreviation	Description
SW	FASCAM, wide-area mine
	UXO Hazards
UB	UXO hazard, bomb (air delivered)
UC	UXO hazard, cluster bomb units
UD	UXO hazard, dual-purpose improved conventional munitions (DPICM)
UG	UXO hazard, guided missile
UJ	UXO hazard, submunitions
UM	UXO hazard, MLRS
UP	UXO hazard, projectile
UR	UXO hazard, dispenser
UT	UXO hazard, thrown ordnance or grenade
UX	UXO hazard, unidentified
	Wire Obstacles
WA	Wire obstacle, double apron
WB	Wire obstacle with booby traps
WF	Wire obstacle, tanglefoot
WG	Wire obstacle, general purpose barbed-taped obstacle (GPBTO)
WN	Wire obstacle, nonstandard
WR	Wire obstacle, roadblock
WT	Wire obstacle, triple standard

g. **Characters 10 and 11: Obstacle/Hazard Number.** Number obstacles and hazards in the order planned or report received (respectively). Numbering starts with "1". If the headquarters needs additional numbers, simply use a 12th character for "100." The 77th Infantry Division receives a report of an unidentified UXO hazard by a subordinate unit. The hazard is the 23rd obstacle/hazard planned or reported in the division AO. Using the previous scenario, this hazard is shown as I077-B2A-UX-23.

h. **Character 12: Status.** The final character of the numbering system provides the status of the obstacle or hazard. The status, using the four symbols at Table F-1, is either planned, being prepared but not executed, or executed/completed.

Appendix G
UXO Training

1. Background

At a minimum, unit commanders should ensure their forces receive training on identification of UXO and the proper conduct of operations around UXO. This appendix provides an example of basic skills that personnel should employ to improve operational safety in the proximity of UXO. These skills serve as guidelines for UXO training—they are not intended to detract from a unit's specific training requirements. Units can receive additional training from combat engineer and EOD personnel. For information pertaining to mines/minefields, contact a combat engineer unit. For additional information on UXO, contact an EOD unit.

2. Training Curriculum

a. **Introduction.** UXO is inherently hazardous! The deadly capabilities of UXO are categorized as conventional, chemical, biological, or any combination thereof. They pose a risk of injury or death to all personnel. They take the form of missiles, bombs, rockets, mines or other devices and can range in size from very small to extremely large.

b. **Training.** If a UXO hazard, or object suspected to be unexploded ordnance is discovered, there are four important steps to take: identify, mark, evacuate, and report.

(1) *Identify.* See Appendix A for additional guidance. When a UXO hazard (mines included) is discovered, proper identification is important to report the hazard accurately. Accurate reporting of the hazard assists the experts in determining clearance requirements and accurately warning friendly units of the hazard. Take the following steps upon recognizing a UXO hazard:

(a) Identify the type UXO, if possible (known ordnance).

(b) Note features (categories such as dropped, projected, thrown or placed) and size, shape, color, and condition of the ordnance (intact or leaking).

(2) *Mark.* Marking the hazard warns other friendly personnel and provides a locating mechanism for clearance by qualified personnel. See Appendix B for additional guidance. Take the following steps when marking UXO:

(a) Mark the hazard from where you are. Do not approach any closer to the hazard.

(b) Use the standard UXO or mine markers; if standard markers are not available, use whatever material (such as engineer tape or chemical lights) available.

(c) Ensure markings are visible in all directions, even at night.

(3) *Evacuate.* Take the following steps:

(a) Evacuate all personnel and equipment from the area.

(b) If evacuation is impossible, isolate or barricade the area. See Appendix D for additional information on isolation and barricade techniques.

(4) *Report.* Take the following steps to report a UXO hazard:

(a) Report the incident using the nine-line format at Appendix C.

(b) If sending the report by radio, transmit from a minimum-safe distance (100 meters) to avoid detonation. Reporting by radio can send an electrical pulse to the hazard, causing detonation.

Provide all pertinent details: size, shape, color, condition, landmarks, and grid coordinates.

REFERENCES

NATO

STANAG 2002, Warning Signs for the Marking of Contaminated or Dangerous Land Areas, Complete Equipment, Supplies, and Stores, Edition 7

STANAG 2079, *Rear Area Security and Rear Area Damage Control*, Edition 4

STANAG 2143, Explosive Ordnance Reconnaissance/Explosive Ordnance Disposal, Edition 4

STANAG 2369, *Identification and Disposal of Surface and Air Munitions*, Edition 2

STANAG 2371, *Explosive Ordnance Disposal Radiography*, Edition 1

STANAG 2377, *Procedures for the Management of an EOD Incident for use when Working with Other Agencies*, Edition 1

STANAG 2389, *Minimum Standards of Proficiency for Trained Explosive Ordnance Disposal Personnel*, Edition 1

STANAG 2391, *EOD Recovery Operations on Fixed Installations*, Edition

STANAG 2485, *Engineer Countermine Operations*

STANAG 2834, *The Operation of the Explosive Ordnance Disposal Technical Information Center*, Edition 2

STANAG 2884, *Underwater Munitions Disposal Procedures*, Edition 2

STANAG 2897, *Standardization of EOD Equipment Requirements and Equipment*, Edition 3

STANAG 2929, *Airfield Damage Repair*, Edition 4

STANAG 2957, *International System used by Armed Forces in the Nuclear Field*, Edition 3

STANAG 2991 (AAP-19C), *NATO Combat Engineer Glossary*

Joint

JP 1-02, *Department of Defense Dictionary of Military and Associated Terms,* 24 January 2000

JP 3-0, *Doctrine for Joint Operations*, 1 February 1995

JP 3-07, *Joint Doctrine for Military Operations Other than War*, 16 June 1995

JP 3-07.5, *Joint Tactics, Techniques, and Procedures for Non-Combatant Evacuation Operations,* 30 September 1997

JP 3-15, *Joint Doctrine for Barriers, Obstacles, and Mine Warfare*, 24 February 1999

JP 3-34, *Engineer Doctrine for Joint Operations*, 05 July 2000

JP 4-04, *Joint Doctrine for Civil Engineering Support*, 26 September 1995

JP 5-0, *Doctrine for Planning Joint Operations*, 13 April 1995

JP 5-00.2, *Joint Task Force Planning Guidance and Procedures*, 13 January 1999

Multiservice

Ordnance Data (ORDATA) Series on compact disk, *Enhanced International Deminer's Guide to UXO Identification*, no date. Obtain disks (need to know) at e-mail: Ordata@eodpoe2.navsea.navy.mil

Technical Manual (TM) (Army), Technical Order (TO) (Air Force), EOD Bulletin (Navy), 60A-1-1-7, *Explosive Ordnance Disposal Procedures; Field Evaluation and Intelligence*, 23 March 1999

AR 75-14/OPNAVINST 8027.1G/MCO 8027.1D/AFJI 32-3002, *Interservice Responsibilities and Procedures for Explosive Ordnance Disposal*, 14 February 1992

FM 4-30.16/MCRP 4-5.1/NTTP 3-02.5/AFTTP(I) 3-2-12, *Multiservice Procedures for Explosive Ordnance Disposal in a Joint Environment*, February 2000

Army

Note. The Army's new numbering system for Field Manuals is the first number listed below, followed by the old FM number.

AR 75-15, *Responsibilities and Procedures for Explosive Ordnance Disposal,* 1 November, 1978

FM 3-0.1 (FM 101-5-1), *Operational Terms and Graphics*, 30 September 1997

FM 3-24.32 (FM 20-32), *Mine/Countermine Operations*, 29 May 1998

FM 3-34 (FM 5-100), *Engineer Operations*, 27 February 1996

FM 3-34.1 (FM 90-7), *Combined Arms Obstacle Integration*, 29 September 1994

FM 3-34.2 (FM 90-13-1), *Combined Arms Breaching Operations*, 28 February 1991

FM 3-34.211 (FM 5-116), *Engineer Operations Echelons Above Corps*, 9 February 1999

FM 3-34.225 (FM 5-71-100), *Division Engineer Combat Operations*, 22 April 1993

FM 3-34.226 (FM 5-100-15), *Corps Engineer Operations*, 6 June 1995

FM 4-30.11 (FM 21-16), *Unexploded Ordnance Procedures*, 30 August 1994

FM 4-30.12 (FM 9-15), *Explosive Ordnance Disposal Service and Unit Operations*, 8 May 1996

Technical Manual (TM) 43-0001-28, *Army Ammunition Data Sheets for Artillery Ammunitions: Guns, Howitzers, Mortars, Recoilless Rifles, Grenade Launchers, and Artillery Fuses*, 31 July 1996

TM 43-0001-29, *Army Ammunition Data Sheets for Grenades*, 30 June 1994

TM 43-0001-36, *Army Ammunition Data Sheets for Land Mines*, 1 September 1994

TM 43-0001-37, *Army Ammunition Data Sheets: Military Pyrotechnics*, 6 January 1994

Graphic Training Aid (GTA) 9-12-1, *Unexploded Ordnance (UXO) Procedures*, 3 January 1992

Marine Corps

MCRP 3-17.2A, *Unexploded Ordnance Procedures*, 30 August 1994

MCWP 3-2, *Aviation Operations*, April 1999

MCWP 3-16, *Techniques and Procedures for Fire Support Planning/Coordination*, March 1992

MCWP 3-17.2, *Explosive Ordnance Disposal*, December 1993

MCWP 4-11.5, *Seabee Operations in the (MAGTF) Marine Amphibious Group Task Force*, November 1997

MCO 3571.2, *Explosive Ordnance Disposal Program*, August 1990

Navy

Department of the Navy, *US Navy Explosive Ordnance Disposal (EOD) Plan*, 28 March 1997

OPNAVINST 3501.93D, *Projected Operational Environment and Required Operational Capabilities for Naval Beach Groups and Their Elements*, 29 August 1994

OPNAVINST 3501.97, *Projected Operational Environment and Required Operational Capabilities for Explosive Ordnance Disposal Ground Forces,* 26 January 1996

OPNAVINST 5450.46K, *Naval Construction Force Policy*, 25 May 1999

NWP 3-02.4, *Explosive Ordnance Disposal*, July 1997

NWP 4-04.1, *Seabee Operations in the (MAGTF) Marine Amphibious Group Task Force,* November 1997

Navy Supplemental Publication 409 MILSTRIP/MILSTRAP Desk Guide, Revised June 1998

US Naval Sea Systems Command, *US Naval Mine Familiarization*, January, 2000

Air Force

AF Policy Document (AFPD) 32-30, *Explosive Ordnance Disposal,* 20 July 1994

AF Instruction (AFI) 32-3001, *EOD Program*, 1 June 1998

AF Instruction (AFI) 33-360, *Publications Management Program*, 31 Jul 98

AF Manual (AF MAN) 10-100, *Airman's Manual,* 1 Aug 1999

AF Pamphlet (AF PAM) 10-219, *Volume 2, Pre-Attack and Disaster Preparations*, 1 November 1996

AF PAM 10-219, *Volume 3, Post-Attack and Post-Disaster Procedures*, 1 April 1997

Air Force Handbook (AFH) 32-4014, *Volume 4, USAF Ability To Survive And Operate Procedures In A Nuclear, Biological, and Chemical (NBC) Environment,* 1 March 1998

Air Force Visual Aid (Afva) 32-4022, *Usaf Unexploded Explosive Ordnance Recognition And Reporting Chart*, 1 April 1999

Glossary

PART I—ABBREVIATIONS AND ACRONYMS

A

AB	rubble by blade (abbreviation used on an engineer obstacle overlay)
ACB	amphibious construction battalion
AD	antitank ditch (abbreviation used on an engineer obstacle overlay)
ADAM	area-denial artillery munitions
ADCON	administrative control
AE	rubble by explosives (abbreviation used on an engineer obstacle overlay)
AF	Air Force
AFB	Air Force base
AFH	Air Force Handbook
AFI	Air Force Instruction
AF MAN	Air Force Manual
AFPD	Air Force Policy Document
AFTTP(I)	Air Force tactics, techniques, and procedures (interim)
AFVA	Air Force Visual Aid
AH	log hurdles (abbreviation used on an engineer obstacle overlay)
AL	1. Alabama; 2. Log crib, log obstacles (abbreviation used on an engineer obstacle overlay)
ALSA	Air Land Sea Application
AM	moveable military operations on urbanized terrain (MOUT) obstacle (as an example a car or a bus)—(abbreviation used on an engineer obstacle overlay)
AMAT	anti-materiel
AN	expedient obstacle, nonstandard (abbreviation used on an engineer obstacle overlay)
AO	area of operations
AP	post obstacles (abbreviation used on an engineer obstacle overlay)
APERS	antipersonnel
AR	Army Regulation
ARFOR	Army forces
AT	antitank
ATTN	attention
AZ	Arizona

B

BA	bridge demolition, abutment (abbreviation used on an engineer obstacle overlay)
BC	bridge demolition, combination of abutment and span (abbreviation used on an engineer obstacle overlay)
BEEF	base emergency engineering force
BLU	bomb live unit
BS	bridge demolition, span (abbreviation used on an engineer obstacle overlay)

C

C2	command and control
CA	California
CBU	cluster bomb unit
CE	civil engineer
CO	Colorado
COMM	commercial phone number
CONPLAN	contingency plan
CVBG	carrier battle group

D

DA	Department of the Army
D.C.	District of Columbia
DOD	Department of Defense
DP	dual-purpose
DPICM	dual-purpose improved conventional munitions
DS	direct support
DSN	Defense Switched Network
DTG	date time group

E

E-mail	electronic mail
EB	enemy bridge demolition (abbreviation used on an engineer obstacle overlay)
ED	enemy antitank ditch (abbreviation used on an engineer obstacle overlay)
EE	enemy rubble obstacle (abbreviation used on an engineer obstacle overlay)
EL	enemy log obstacle (abbreviation used on an engineer obstacle overlay)
EM	enemy minefield (abbreviation used on an engineer obstacle overlay)
ENY	enemy-generated UXO (abbreviation used for graphic identification on an engineer obstacle overlay)
EOD	explosive ordnance disposal

ER	enemy road crater (abbreviation used on an engineer obstacle overlay)
ES	enemy scatterable minefield (abbreviation used on an engineer obstacle overlay)
ET	enemy abatis (abbreviation used on an engineer obstacle overlay)
EW	enemy wire obstacle (abbreviation used on an engineer obstacle overlay)
EX	enemy unidentified hazard (abbreviation used on an engineer obstacle overlay)

F

FASCAM	family of scatterable mines
FL	Florida
FM	field manual (Army)
FSSG	force service support group

G

G-4	Army or Marine Corps component logistics staff officer
GA	Georgia
GEMSS	ground-emplaced, mine-scattering system
GP	general purpose
GPBTO	general-purpose barbed-tape obstacle
GS	general support

H

HE	high explosive
HI	Hawaii
HQ	headquarters

I

ICM	improved conventional munitions
IED	improvised explosive device

J

J-2	Intelligence Directorate of a joint staff
J-3	Operations Directorate of a joint staff
J-4	Logistics Directorate of a joint staff
J-5	Plans Directorate of a joint staff
JEODOC	joint explosive ordnance disposal operations center
JEODTF	joint explosive ordnance disposal task force
JFC	joint force commander
JOA	joint operations area
JP	joint publication
JRA	joint rear area
JRAC	joint rear area coordinator
JTF	joint task force

L

LOC	lines of communications
LNO	liaison officer

M

MAGTF	Marine air-ground task force
MB	minefield, block (abbreviation used on an engineer obstacle overlay)
MCO	Marine Corps Order
MCRP	Marine Corps Reference Publication
MCWP	Marine Corps Warfare Publication
MD	1. Maryland; 2. minefield, disrupt (abbreviation used on an engineer obstacle overlay)
MEF	Marine expeditionary force
METT-T	mission, enemy, terrain and weather, troops and support available, time available (DOD)
METT-TC	mission, enemy, terrain and weather, troops and support available, time available, and civilian considerations (Army only)
MF	minefield, fix (abbreviation used on an engineer obstacle overlay)
MILSTRIP	military standard requisitioning and issue procedure
MLRS	multiple launch rocket system
MN	minefield, nonstandard pattern (abbreviation used on an engineer obstacle overlay)
MO	Missouri
MOOTW	military operations other than war
MOPMS	modular pack mine system
MOUT	military operations on urban terrain
MP	minefield, protective (abbreviation used on an engineer obstacle overlay)
MQ	minefield, nuisance (abbreviation used on an engineer obstacle overlay)
MS	minefield, standard pattern (abbreviation used on an engineer obstacle overlay)
MT	minefield, turn (abbreviation used on an engineer obstacle overlay)
MTTP	multiservice tactics, techniques, and procedures

N

NATO	North Atlantic Treaty Organization
NBG	Naval beach group
NBC	nuclear, biological, chemical
NC	North Carolina
NCB	Naval construction brigades

NCG	Naval construction groups
NCF	Naval construction force
NCR	Naval construction regiment
NSN	national stock number
NTTP	Navy tactics, techniques, and procedures
NWDC	Navy Warfare Development Command
NWP	naval warfare publication

O

OD	olive drab
OH	Ohio
OPCOM	operational command (NATO)
OPCON	operational control
OPLAN	operation plan
OPNAVINST	operational Naval instruction
OPORD	operations order

P

PDM	pursuit deterrent munitions

R

RAAM	remote anti-armor mine system
RD	road crater, deliberate (abbreviation used on an engineer obstacle overlay)
RED HORSE	**Rapid Engineers Deployable Heavy Operational Repair Squadron, Engineers**
RH	road crater, hasty (abbreviation used on an engineer obstacle overlay)
RI	Rhode Island
RM	road crater, mined (abbreviation used on an engineer obstacle overlay)

S

SA	family of scatterable mines (FASCAM), area-denial artillery munitions (ADAM)—(abbreviation used on an engineer obstacle overlay)
SB	FASCAM, Gator (abbreviation used on an engineer obstacle overlay)
SF	FASCAM, ADAM or RAAM (abbreviation used on an engineer obstacle overlay)
SG	FASCAM, ground-emplaced, mine-scattering system (GEMMS) (abbreviation used on an engineer obstacle overlay)
SM	FASCAM, MOPMS (abbreviation used on an engineer obstacle overlay)

SP	FASCAM, pursuit deterrent munitions (PDM) (abbreviation used on an engineer obstacle overlay)
SR	FASCAM, RAAM (abbreviation used on an engineer obstacle overlay)
STANAG	standardization agreements
SV	FASCAM, Volcano (abbreviation used on an engineer obstacle overlay)
SW	FASCAM, wide-area mine (WAM)

T

TAA	tactical assemble area, tanker alert area, total Army authorization, training advisory alert
TACON	tactical control
TF	task force
TM	technical manual
TOW	tube-launched, optically tracked, wire-guided
TPFDD	time-phased force and deployment data
TTP	tactics, techniques, and procedures
TX	Texas

U

UB	UXO hazard, bomb (air delivered)—(abbreviation used on an engineer obstacle overlay)
UC	UXO hazard, cluster bomb unit (abbreviation used on an engineer obstacle overlay)
UD	UXO hazard, DPICM (abbreviation used on an engineer obstacle overlay)
UG	UXO hazard, guided missile (abbreviation used on an engineer obstacle overlay)
UIC	unit identification code
UJ	UXO hazard, submunitions (abbreviation used on an engineer obstacle overlay)
UM	UXO hazard, MLRS (abbreviation used on an engineer obstacle overlay)
UP	UXO hazard, projectile (abbreviation used on an engineer obstacle overlay)
UR	UXO hazard, dispenser (abbreviation used on an engineer obstacle overlay)
U.S.	United States
USAF	United States Air Force
USMC	United States Marine Corps
USN	United States Navy
UT	1. Utah; 2. UXO hazard, thrown ordnance or grenade (abbreviation used on an engineer obstacle overlay)
UTC	unit type codes

UX	UXO hazard, unidentified (abbreviation used on an engineer obstacle overlay)
UXO	unexploded explosive ordnance

V

VA	Virginia
VSW	very shallow water

W

WA	wire obstacle, double apron (abbreviation used on an engineer obstacle overlay)
WB	wire obstacle, with booby traps (abbreviation used on an engineer obstacle overlay)
WF	wire obstacle, tanglefoot (abbreviation used on an engineer obstacle overlay)
WG	wire obstacle, (abbreviation used on an engineer obstacle overlay)
WN	wire obstacle, nonstandard (abbreviation used on an engineer obstacle overlay)
WR	wire obstacle, roadblock (abbreviation used on an engineer obstacle overlay)
WT	wire obstacle, triple standard (abbreviation used on an engineer obstacle overlay)

PART II—TERMS AND DEFINITIONS

abatis. A vehicular obstacle constructed by felling trees (leaving a 1- to 2- meter stump above the ground on both sides of a road, trail, gap, or defile) so they fall interlocked, toward the expected direction of enemy approach. The trees should remain attached to the stumps and be at a 45-degree angle to the roadway. The obstacle itself should be at least 75 meters in depth to be most effective.

administrative control (ADCON). Direction or exercise of authority over subordinate or other organizations in respect to administration and support, including organization of service forces, control of resources and equipment, personnel management, unit logistics, individual and unit training, readiness, mobilization, demobilization, discipline, and other matters not included in the operational missions of the subordinate or other organizations.

anti-disturbance fusing. See fuse, anti-disturbance.

anti-fratricide. Actions taken to eliminate or reduce casualties of friendly forces by their own ordnance or equipment.

area of operations (AO). An operational area defined by the joint force commander for land and naval forces. Areas of operation do not typically encompass the entire operational area of the joint force commander, but should be large enough for component commanders to accomplish their missions and protect their forces.

barricade. A structure composed essentially of concrete, earth, metal, or wood, or any combination thereof, and so constructed as to reduce or confine the blast effect and fragmentation of an explosion.

battlespace. The environment, factors, and conditions, which must be understood to successfully apply combat power, protect the force, or complete the mission. This includes the air, land, sea, space, and the included enemy and friendly forces, facilities, weather, terrain, the electromagnetic spectrum, and information environment within the operational areas and areas of interest.

blast. The brief and rapid movement of air, vapor or fluid away from a center of outward pressure, as in an explosion or in the combustion of rocket fuel: the pressure accompanying this movement. This term is commonly used for "explosion," but the two terms may be distinguished.

bomblet. See submunition.

booby trap. An explosive or non-explosive device or other item, deliberately placed to cause casualties when an unsuspecting person disturbs an apparently harmless object or performs a normally safe act.

breaching. A tactical task where any means available is employed to break through or secure a passage through an enemy defense, obstacle, minefield, or fortification. (FM 101-5-1)

carrier battle group. (DOD) A standing naval task group consisting of a carrier, surface combatants, and submarines as assigned in direct support, operating in mutual support with the task of destroying hostile submarine, surface, and air forces within the group's assigned operational area and striking at targets along hostile shore lines or projecting fire power inland.

clearance operations. In countermine operations, actions to restore the freedom of movement within a mined area or along a mined route.

clearance by priority area. Clearing areas critical to mission essential operations by using a priority list.

cluster bomb unit. An aircraft store composed of a dispenser and submunitions.

combat support. Fire support and operational assistance provided to combat elements.

combat service support. (DOD) The essential capabilities, functions, activities, and tasks necessary to sustain all elements of operating forces in theater at all levels of war. Within the national and theater logistic systems, it includes but is not limited to that support rendered by service forces in ensuring the aspects of supply, maintenance, transportation, health services, and other services required by aviation and ground combat troops to permit those units to accomplish their missions in combat. Combat service support encompasses those activities at all levels of war that produce sustainment to all operating forces on the battlefield.

command and control. (DOD) The exercise of authority and direction by a properly designated commander over assigned and attached forces in the accomplishment of the mission. Command and control functions are performed through an arrangement of personnel, equipment, communications, facilities, and procedures employed by a commander in planning, directing, coordinating, and controlling forces and operations in the accomplishment of the mission. Also called C2.

communications zone. Rear part of theater of operations which contains the lines of communications, establishments for supply or evacuation, and other agencies required for the immediate support and maintenance or field forces.

countermine operations. In land mine warfare, an operation to reduce or eliminate the effects of mines and minefields.

counter-mobility. Limiting the enemies' ability to maneuver its forces.

defilade. (DOD, NATO) 1. Protection from hostile observation and fire provided by an obstacle such as a hill, ridge, or bank. 2. A vertical distance by which a position is concealed from enemy observation. 3. To shield from enemy fire or observation by using natural or artificial obstacles.

diagnostic. Those actions taken to identify and evaluate unexploded explosive ordnance.

dispenser. An item designed to be mounted, but not permanently fixed, on aircraft to carry and eject small ordnance.

explosive ordnance. All munitions containing explosives, nuclear fission or fusion materials and biological and chemical agents. This includes bombs and warheads; guided and ballistic missiles; artillery, mortar, rocket, and small arms ammunition; all mines, torpedoes, and depth charges; demolition charges; pyrotechnics; clusters and dispensers; cartridge and propellant actuated devices; electro-explosive devices; clandestine and improvised explosive devices; and all similar or related items or components explosive in nature.

explosive ordnance disposal (EOD). The detection, identification, on-site evaluation, rendering safe, recovery, and final disposal of unexploded explosive ordnance. It may also include explosive ordnance that has become hazardous by damage or deterioration.

forward arming and refueling point. A temporary facility, organized, equipped, and deployed by an aviation commander, and normally located in the main battle area closer to the area of operation than the aviation unit's combat service area, to provide fuel and ammunition necessary for the employment of aviation maneuver units in combat. The forward arming and refueling point permits combat aircraft to rapidly refuel and rearm simultaneously.

fratricide. Killing of friendly forces by U.S. fires or equipment.

fuse, anti-disturbance. A fuse designed to become armed after impact, or after being emplaced, so that any further movement or disturbance will result in detonation.

fuse, proximity. A fuse wherein primary initiation occurs by remotely sensing the presence, distance, and/or direction of the target through the characteristics of the target itself or its environment.

high water mark. Properly, a mark left on a beach by wave wash at the preceding line. Because it can be determined by simple observation, it is frequently used in place of the high-water line, which can be determined only by survey. When so used, it is called the high-water line.

improvised explosive device (IED). A device placed or fabricated in an improvised manner incorporating destructive, lethal, noxious, pyrotechnic, or incendiary chemicals and designed to destroy, incapacitate, harass, or distract. It may incorporate military stores, but is normally devised from nonmilitary components.

joint force commander (JFC). A general term applied to a combatant commander, subunified commander, or joint task force commander authorized to exercise combatant command (command authority) or operational control over a joint force.

joint operations. A general term to describe military actions conducted by joint forces (two or more services), or by service forces in relationships (for instance support or coordinating authority), which, of themselves, do not create joint forces.

joint operations area. An area of land, sea, and airspace, defined by joint forces, or by service forces in relationships (such as support, coordinating authority), which, of themselves, do not create joint forces. Also called JOA.

lines of communications (LOC). A route, either land, water, and/or air, which connects an operating military force with a base of operations and along which supplies and military forces move.

Marine air-ground task force (MAGTF). A task organization of Marine forces (division, aircraft wing, and service support groups) under a single command and structured to accomplish a specific mission. The MAGTF components will normally include command, aviation combat, ground combat, and combat service support elements (including Navy Support Elements). Three types of MAGTF can be task organized: Marine expeditionary unit, Marine expeditionary brigade, and Marine expeditionary force.

military operations other than war (MOOTW). Operations that encompass the use of military capabilities across the range of military operations short of war. These military actions can be applied to complement any combination of the other instruments of national power and occur before, during, and after war.

mine. (1) In land mine warfare, an explosive or other material, normally encased, designed to destroy or damage ground vehicles, boats, or aircraft, or designed to wound, kill, or otherwise incapacitate personnel. It may be detonated by the action of its victim, by the passage of time, or by controlled means. (2) In naval mine warfare, an explosive device laid in the water with the intention of damaging or sinking ships or of deterring shipping from entering an area. The term does not include devices attached to the bottoms of ships or to harbor installations by personnel operating underwater, nor does it include devices that explode immediately on expiration of a predetermined time after laying.

minefield. A space either on land or in water in which mines have been placed with or without pattern. Usually the mines are related one to another in such a manner as to collectively achieve the purpose for which the field was established.

mobility. (DOD, NATO) A quality or capability of military forces which permits them to move from place to place while retaining the ability to fulfill their primary mission.

munitions. A complete device charged with explosives, propellants, pyrotechnics, initiating composition, or nuclear, biological, or chemical material for use in military operations, including demolitions. Certain suitably modified munitions can be used for training, ceremonial or non-operational purposes. Also called ammunition. (**Note.** In common usage, "munitions" (plural) can be military weapons, ammunition, and equipment.)

nap of the earth. (DOD, NATO) Flight close to the Earth's surface during which airspeed, height and/or altitude are adapted to the contours and cover of the ground in order to avoid enemy detection and fire.

obstacle. Any obstruction designed or employed to disrupt, fix, turn, or block the movement of an opposing force, and to impose additional losses in personnel, time, and equipment on the opposing force. Obstacles can exist naturally or can be manmade, or can be a combination of both.

operational control (OPCON). Transferable command authority that may be exercised by commanders at any echelon at or below the level of combatant command. OPCON is inherent in combatant command (command authority). OPCON may be delegated and is the authority to perform those functions of command over subordinate forces involving organizing and employing commands and forces, assigning tasks, designating objectives, and giving authoritative direction necessary to accomplish the mission. OPCON includes authoritative direction over all aspects of military operations and joint training necessary to accomplish missions assigned to the command. OPCON should be exercised through the commanders of subordinate organizations. Normally this authority is exercised through subordinate joint force commanders and service and/or functional component commanders. OPCON normally provides full authority to organize commands and forces and to employ those forces as the commander in operational control considers necessary to accomplish assigned missions. OPCON does not, in and of itself, include authoritative direction for logistics or matters of administration, discipline, internal organization, or unit training.

operation plan. (DOD) Any plan, except for the Single Integrated Operation Plan, for the conduct of military operations. Plans are prepared by combatant commanders in response to requirements established by the Chairman of the Joint Chiefs of Staff and by commanders of subordinate commands in response to requirements tasked by the establishing unified commander. Operation plans are prepared in either a complete format (OPLAN) or as a concept plan (CONPLAN). The CONPLAN can be published with or without a time-phased force and deployment data (TPFDD) file.

a. OPLAN—An operation plan for the conduct of joint operations that can be used as a basis for development of an operation order (OPORD). An OPLAN identifies the forces and supplies required to execute the CINC's Strategic Concept and a movement schedule of these resources to the theater of operations. The forces and supplies are identified in TPFDD files. OPLANs will include all phases of the tasked operation. The plan is prepared with the appropriate annexes, appendixes, and TPFDD files as described in the Joint Operation Planning and Execution System manuals containing planning policies, procedures, and formats. Also called OPLAN.

b. CONPLAN—An operation plan in an abbreviated format that would require considerable expansion or alteration to convert it into an OPLAN or OPORD. A CONPLAN contains the CINC's Strategic Concept and those annexes and appendixes deemed necessary by the combatant commander to complete planning. Generally, detailed support requirements are not calculated and TPFDD files are not prepared. Also called CONPLAN.

c. CONPLAN with TPFDD—A CONPLAN with TPFDD is the same as a CONPLAN except that it requires more detailed planning for phased deployment of forces.

operation order. (DOD) A directive issued by a commander to subordinate commanders for the purpose of effecting the coordinated execution of an operation. Also called OPORD.

ordnance. Explosives, chemicals, pyrotechnics, and similar stores—to include bombs, guns and ammunition, flares, smoke, and napalm.

protective barricade. Barricade constructed around personnel, equipment, or resources to shelter from overpressure and impact of high-angle low-speed fragments of exploding ordnance.

proximity fusing. See fuse, proximity.

recovery. In naval mine warfare, salvage of a mine as nearly intact as possible to permit further investigation for intelligence and/or evaluation purposes.

render safe. As applied to weapons and ammunition, the changing from a state of readiness for initiation to a safe condition.

scatterable mines. In land mine warfare, a mine laid without regard to classical pattern and designed to be delivered by aircraft, artillery, missile, ground dispenser or hand. Once laid it normally has a limited laid life.

self-extract. Action of removing personnel and equipment from a hazard area.

service component command. A command consisting of the service component commander and all those service forces, such as individuals, units, detachments, organizations, and installations under the command, including the support forces that have been assigned to a combatant command, or further assigned to a subordinate unified command or joint task force.

special operations forces. (DOD) Those active and reserve component forces of the military services designated by the Secretary of Defense and specifically organized, trained, and equipped to conduct and support special operations. Also called SOF.

submunition. Any munition that, to perform its task, separates from a parent munition.

suppressive barricade. Barricade constructed to isolate an explosion, to deflect the thermal/shockwave and absorb low angle-high speed fragments of exploding ordnance.

unexploded explosive ordnance (UXO). Explosive ordnance that has been primed, fused, armed or otherwise prepared for action, and fired, dropped, launched, projected or placed in such a manner as to constitute a hazard to operations, installations, personnel or materiel and remains unexploded either by malfunction or design or for any other cause.

UXO hazard area. Area that is contaminated with unexploded explosive ordnance to include the blast and fragmentation radius of the munitions item.

UXO incident. Situation that requires an EOD response and/or procedures.

unit type code (UTC). A five-character, alphanumeric code that uniquely identifies each type unit, and in the case of this manual specific EOD organizations or structures of the Armed Forces.

weapons of mass destruction. In arms control usage, weapons that are capable of a high order of destruction and/or of being used in such a manner as to destroy large numbers of people. Can be nuclear, chemical, biological, and radiological weapons, but excludes the means of transporting or propelling the weapon where such means is a separable and divisible part of the weapon.

Index

E

F

G

H

I

J

L

M

N

O

P

R

S

T

U

V

W

Made in United States
Troutdale, OR
11/03/2026

56844392R00044